T0104422

OFF THE HOOK

NEVILLE WILLIAMS

Order this book online at www.trafford.com
or email orders@trafford.com

Most Trafford titles are also available at major online book retailers.

© Copyright 2017 Neville Williams.
All rights reserved. No part of this publication may be reproduced, stored in a
retrieval system, or transmitted, in any form or by any means, electronic, mechanical,
photocopying, recording, or otherwise, without the written prior permission of the author.

Print information available on the last page.

ISBN: 978-1-4907-7944-7 (sc)
ISBN: 978-1-4907-7943-0 (e)

Because of the dynamic nature of the Internet, any web addresses or links contained in
this book may have changed since publication and may no longer be valid. The views
expressed in this work are solely those of the author and do not necessarily reflect the
views of the publisher, and the publisher hereby disclaims any responsibility for them.

Any people depicted in stock imagery provided by Thinkstock are models,
and such images are being used for illustrative purposes only.
Certain stock imagery © Thinkstock.

Trafford rev. 12/16/2016

 www.trafford.com

North America & international
toll-free: 1 888 232 4444 (USA & Canada)
fax: 812 355 4082

Contents

Preface

It seems in the present age, the only people speaking–out against evolution are those who are already confused by their own belief system.

False religions, by opposing the views of Darwin are merely trying to bolster their own "handed down" traditions. My approach to evolution is to start with the assumption that Charles Darwin was a genuine scientist and because of that he considered carefully and wrote down his theory only after thinking about what he was writing and (more importantly) how it would be used.

In my first book you will see that science opposes the views of fundament religion. (And they have a right to feel that way). If you have any doubt that I might be a "churchy" or religious person then you could read my first book - as an introduction to this work. In the book "Artificial Religion,"

I enjoyed writing about the work of Richard Dawkins as he clearly shows that the various religions of this world are man–made. The religions are simply too disordered to be considered genuine in any way.

I am neither a religionist, nor an atheist––that's what makes this book unique. The reader will find that the scientific work of Charles Darwin is covered in an unusual way, just as the churchy ideas were covered in an unusual way in the accurate account of false religion, produced by Richard Dawkins (known as "The Virus of faith").

In *this* book you will see some examples of the early work of Charles Darwin, perhaps things that were not taught in "science" at your local school.

Before I move on to the real DARWIN, another matter could be cleared up, that is the strange practice of *'guilt by association.'* For example, if I use material from a particular book, magazine or web site, I will not have that work dismissed just because someone doesn't like the *source* of the information.

I may use material from a person who 'claims' to be a 'young earth scientist' for example, that doesn't mean I support that particular belief system; and even more importantly... that doesn't mean that the information shown is defective in some way. Personal opinion cannot be used as a part of science, (that would make it nonsense). If there really is factual information present, why be concerned where the facts come from? Deal with the information. Either the information is true, or it is not. Attacking the source of the information is a futile exercise and it must **never** be a preferred method of defending science.

The 'young earth science' is simply following a tradition that has been handed down, rather than taking the time to consider truthful information. Sure some parts of their work are accurate, but they do not follow the book that they claim to follow. They are just another branch of false religion. If you happen to feel "uncomfortable" with any of the facts that are presented, then you still must either agree or disagree – BASED ON THE FACTS.

Don't think that you can dismiss my work simply by chanting, "he must be one of those." I accept that most religious folk are on the wrong track, but that doesn't mean that ALL of their ideas are false. Some of the information they present is truthful and accurate and deserves your attention!

The method of dismissing work from sites other than those acceptable to science is part of the way that a scientist is trained. It is definitely not a natural way to behave. Facts are facts regardless of your 'feelings.' All things in science must be either proved or disproved by careful consideration. Anyone who goes for instant–dismissal without thought is a person dealing in nonsense.

If I hear a commentator saying "oh he's just one of those," I know straight away, that person is not concerned about right or wrong, he is only concerned with upholding his own personal belief. Simply keep the faith in science and there is no need to study or research.

The problem is, by treading–down any opposition to their personal world–view a scientist is denying himself the right to learn. Science must be about learning new and novel information. A genuine scientist must have the desire to investigate, no matter where the information is heading. A logical mind and careful attention to detail are all that is needed for a change of attitude. In other words, deal with the information. Then either dismiss it or agree with it.

To dismiss information outright, without a second thought, would hardly be considered a scientific approach! It takes a strong person to consider information properly, to consider facts and figures, without prejudice. It would certainly be an error to dismiss an idea simply because it isn't popular. Try to remember, the millions who attend the Universal Church. They do so because it is popular. If you take that as an example, you can easily see that popular doesn't always mean correct. Please remember this when you read about the first person to speak out about the Stars and the Universe. To almost every thinker alive the prospect of an expanding universe (with star–centric systems) seemed absurd.

Science proudly boasts a 97% support for the theory of Darwinism. This book however is about the science of Charles Darwin and the study of scientific evolution. The point I make is that you must consider my work with the idea that it is either true or it is false – rather than attacking my personal character. Who or what I am – has nothing to do with the facts. If the material is logical and factual it will remain the same, regardless of your personal abilities, or my personal abilities. If the material is logical and factual it will remain the same, regardless of the source of the information.

The typical scientist will attempt to win an argument in one of two ways. Either *they* will keep arguing until reaching a point of 'having the last say,' thus believing *they* have won the argument. (That is hardly the scientific approach), or if all else fails *they* will simply say "I am the scientist and you are merely a member of the MAN-KIND, therefore I don't even have to discuss this matter with you."

I say to them... "That method is an arrogant and ignorant way to do business."

That's right *their* chosen method of operation has proven to be the worst form of **intellectual snobbery.**

Chapter One

NATURAL SELECTION

Darwinian Evolution:

As early as 1842 Charles Darwin had the outlines of his theory prepared. Seemingly endless investigation and testing was required, which led to a constant postponing of the final publication.

The idea of 'the evolution of organisms' is very old indeed. It appeared in the work of many Greek philosophers, including Aristotle. However at the time of Darwin's study the idea known as 'fixity of species' was still generally 'preserved' by science.

Darwin spent years of observation and thought (during his journey on the Beagle). Then he further studied and refined his ideas for many years before completing his book. He put a great deal of effort into his work, and we owe it to him to notice and appreciate the *TRUE MEANING* of his thoughts and his observations. This is how Darwin summed up his own work...

> My success as a man of science, whatever this may have amounted to, has been determined, as far as I can judge, by complex and diversified mental qualities and conditions. Of these, the

most important have been—the love of science—unbounded patience in long reflecting over any subject—industry in observing and collecting facts—and a fair share of *INVENTION* as well as of common sense. With such moderate abilities as I possess, it is truly surprising that I should have influenced to a considerable extent the belief of scientific men on some important points.

[This material is from the editor's introduction to ORIGIN OF SPECIES by Charles Darwin]

But the expression often used by Mr. Herbert Spencer of the Survival of the Fittest is more accurate, and is sometimes equally convenient. We have seen that man by selection can certainly produce great results, and can adapt organic beings to his own uses, through the accumulation of slight but useful variations, given to him by the hand of Nature. But **Natural Selection**, as we shall hereafter see, is a power incessantly ready for action, and is immeasurably superior to man's feeble efforts.

But it may be objected that if all organic beings thus tend to rise in the scale, how is it that throughout the world a multitude of the lowest forms still exist? And how is it, that in each great class some forms are far more highly developed than others? Why have not the more highly developed forms everywhere supplanted and exterminated the lower? Lamarck, who believed in an innate and inevitable tendency towards perfection in all organic beings, seems to have felt this difficulty so strongly that he was led to suppose that new and simple 'forms' are continually being produced by spontaneous generation. Science has not as yet proved the truth of this belief, whatever the future may reveal. On our theory the continued existence of lowly organisms offers no difficulty for **natural selection**, or the survival of the fittest does not necessarily include progressive development – it only takes advantage of such variations as arise and are beneficial to each creature under its complex relations of life. [ORIGIN OF SPECIES, Chapter 3, p64]

There are no surprises in the early chapters of "The Origin of Species" Most of the early work includes the stuff that is being taught as [Darwinian] evolution, right up to the present time. For now the early chapters of *his* book can be thought of as a lead-up to the more important matters, such as symbiosis and the fossil record. The key reason for this separation is in the rule of truth. In the early part of my studies, I was taught that NOT TELLING THE WHOLE STORY was exactly the same as telling a LIE.

Some of the material from the first chapters will be covered in another section, and discussed as 'scientific evolution;' but first there needs to be a clear understanding of the vast difference between Darwinian evolution and the *scientific* theory of evolution. Keep in mind that a scientific theory is nothing like a theory in the mind of a normal person. It takes a separate definition to 'beef–up' the value of the word 'theory' for scientific use. You will notice elitist–definitions being used (by science) and the invention of new purpose built words.

In the 1930's there was a well–known saying, "If you're going to tell a lie—make it a big one." New words and new definitions are all a part of the 'science of evolution' and the whole theory collapses without *their* new words and definitions.

It is important to commence with the idea of getting Darwin off the hook. To achieve this, I need to show that there are certain things that Charles Darwin has presented to his readers, which have been ignored by the scientific community.

It is grossly unfair for science to ignore the words of Darwin's book while pinning his name to *their* scientific theory of evolution. To me that sort of attitude is an example of *DOUBLE STANDARDS*.

Double standards can only come from a double–minded person; and a double–minded person is unstable in all his ways.

Before I move on to the clear warnings given by Charles Darwin I need you to be clear about my position.

I am NOT a part of any weird 'man-made' religion or any weird branch of 'science.' I am just a regular person – presenting factual information for you to consider.

In some cases I am passing on the science of Charles Darwin. I accept Charles Darwin as a genuine scientist however I do NOT go to the extreme of celebrating his 200th birthday. In fact I don't make a fuss over the birthday of *any* dead people, scientist or otherwise.

My book takes a very simple approach. I begin by taking a good look at some of the early works of Charles Robert Darwin. I ask the question, would Charles Darwin be accepted by the scientific community of our time? Or to put it another way, would Charles Darwin be allowed to teach 'Darwinian–evolution' to the present day science student?

I also set out to show that there is a glaring difference between a thing called Darwinian evolution and the thing being taught and accepted by science at the present time. That is––the present day teaching which I would call 'the scientific theory of evolution.'

I do understand the difference between a human definition (of a theory) and the upgraded or 'elitist' definition of the word 'theory.' [An elitist is someone who believes in rule by an elite group].

I believe that if Charles Darwin were a part of the scientific community today, he would be labeled a 'crackpot.'

Some readers will not be familiar with the terms 'crackpot' and 'denier,' so I will attempt to explain their use, (as used by the present–day scientific community).

> A 'crackpot' is any person who makes a claim that doesn't fit the mind of the hardened atheist scientist.

A 'denier,' on the other hand, is someone who uses a periodic publication, (or even a web page), to publish the thoughts of any person who has previously been excluded as a 'crackpot.' It's very easy *AND* very common for the scientist to make use of these terms.

When they hear any person, (scientist or otherwise), saying things which do not fit with their personal view, they simply invoke the chant of 'crackpot.' That's all there is to it!

And the following statement is the scientific mumbo–jumbo that allows the scientist the freedom to 'put down' any opposition to a belief in evolution.

"EVEN IF all the data POINTS AWAY FROM OUR THEORY, IT WILL NOT BE ACCEPTED BY SCIENCE, BECAUSE the new data WOULD NOT BE COUNTED AS NATURALISTIC."

Some readers are going to think that I am being a bit 'over the top' with my remarks about the scientific work known as – excluding a crackpot, so I will continue on here with a fine example of a biology student who was rejected by the establishment while actually learning the science of [Darwinian] Evolution.

Silvia Baker:

As I have given the scientific definition of the word 'crackpot' I feel it would be a great benefit to insert a brief section about a real–life crackpot. This insert is included for the benefit of those who would simply dismiss my claim –without a second thought.

Silvia Baker did not set out to oppose the work of scientific–evolution, she merely asked the wrong question while working with, and learning from the hardened evolutionists of her time. To appreciate what Sylvia Baker has to say you need to tie the 'problem' to the work of Charles Darwin. For this you need to consider what Darwin thought when he pondered the science of the human–eye.

"To suppose that the eye with all its inimitable contrivances for adjusting the focus to different distances, for admitting different amounts of light, and for the correction of spherical and chromatic aberration, could have been formed by natural–selection seems, I freely confess, **absurd in the highest degree.**" [Origin of Species –– Difficulties – page 173]

One component of our camera (the eye) is its 'film' or retina, the light–sensitive layer at the back of the eyeball. This layer is thinner than a sheet of Saran Wrap and is vastly more sensitive to light than any man-made film. The best man-made film can handle a range of about 1,000-to one. By comparison, the human retina can handle a dynamic range millions of times greater). The human retina can sense as little as a single photon of light in the dark!

In bright daylight, the retina bleaches out and turns its 'volume control' way down so it won't overload. The light sensitive cells are like complex high–gain amplifiers. There are over 10 million such cells and they are packed into a small space (200,000 per mm), in the highly sensitive Fovea, (the part where vision is most acute). These photoreceptor cells have a very high rate of metabolism and must be completely re–printed and replaced within 7 days.

Darwin couldn't have known about the immense detail inside the eye, but if he had known, it would have been a game–changer for his theory.

After writing about the eye, he then goes on to explain how he might reason through some of the difficulties. Thinking in a "survival" way he would think back to a time before any of the birds—insects—or animals had eyes of any kind.

Each and every part of an evolved [creature] would (by nature) require a time-before; THE EYES ARE NO EXCEPTION.

With that in mind, take a look at the (once keen) student of evolution—Sylvia Baker.

Sylvia was studying at Sussex University in the 1960s. Sylvia Baker was not a radical student. She attended the university as a genuine student of evolutionary biology.

She began as an evolutionist.

While studying A–level zoology, Sylvia realised that the work she was studying was not that convincing, even so she continued her work, still believing that evolution must be true. This was a modern course, in a modern university and students were encouraged to follow the evidence **wherever it may lead.**

The first part of the course had a focus on biochemistry and genetics and similar fields. Towards the end of the course Sylvia Baker got the shock of her life, when she realised that 'science' was being strongly applied to a 'belief' in evolution.

> Sylvia Baker was taught that she had to… follow the evidence wherever it may lead; EXCEPT WHEN THE EVIDENCE DID NOT SUPPORT THE CURRENT PARADIGM.

One day Sylvia attended a seminar where they were discussing the evolution of the human–eye. The way she was thinking may have been something like this… If evolution is true, and if it followed the course that it's supposed to follow;

From molecules, to a simple existence, to cellular creatures, to invertebrates, to vertebrates… then the eyes have to follow that same course! So the whole creature has to evolve from one step to the next or it doesn't fit the evolutionary tree. To think of that another way – you cannot have one part of the "creature" evolving, without the other parts, it has to be all or nothing. The main reason why it must be 'all or nothing' is because only the finished–product would be fit for survival.

To get a clearer idea of what happened (to Silvia Baker) consider the events from *her* point of view…

"One day towards the end of my three years [of the university course]. I was told that we had to follow the evidence wherever it may lead – except where it queried evolution. I was in a seminar where we were discussing the evolution of the vertebrate eye. The eye had to have evolved from an invertebrate type eye to a vertebrate type eye. That was what we were discussing at that seminar.

As we talked round and round in circles on it, well I didn't say much, I just listened to the others. It was clearly impossible because the two systems work on completely different basic principals, so eventually I said, well if we can't see how it's happened, perhaps we should question whether it has happened; and I was absolutely astonished at the response that I got. There was a shocked silence, there had been a lot of discussion and a lot of debate and noise and then the place went completely silent. The person leading the seminar, who until then had been helpful and encouraging, and trying to get us all to say as much as possible, said in a very harsh and decisive way—"I am not going to get involved in any controversy."

So for the very first time, Sylvia Baker was scienced. [You can write your own definition for that word].

The word 'theory' is 'beefed–up' for scientific use. The word evolution is handy because it has two distinctly different meanings, and so on. Any word that science has re–defined is handy for defending or maintaining faith in evolution. What if Charles Darwin himself had been there at the seminar, would he have been scienced also?

The scientific establishment proudly announces (to the world) that 97% of all present–day scientists will support the theory of evolution. What that really means is that there are still some inside the culture of science, who have not yet been crackpotised.

The story of Sylvia Baker was not given here as a scientific work. Yet it is clearly an example of how a person becomes crackpotised... and this can happen even if they are working through a university study of Darwinian evolution. You might ask yourself, would they have behaved that way (towards Sylvia Baker) if Charles Darwin were standing there with her? Would they have allowed Charles Darwin to speak? Would they have allowed *him* to say, "To suppose that the eye with all its inimitable contrivances for adjusting the focus to different distances, for admitting different amounts of light, and for the correction of spherical and chromatic aberration, could have been formed by natural selection, seems, I freely confess, **absurd in the highest degree?"**

And the answer is... I don't think so!

With that in mind, I continue on with the question... "Would the real Charles Darwin be accepted by the scientific community or even the science teachers of our time?" Ask yourself, did Charles Darwin close the door on the study of the human eye... or did he leave the door open, awaiting an explanation?

> What exactly did Darwin write, regarding 'the theory of evolution?' Is there anything (in his work) that scientists would be uncomfortable with? Did Charles Darwin leave important stepping–stones for the scientific community to follow, (such as facts to seek out and verify)? Were there any parts to *his* work that would cause 'Darwinian evolution' to come to a grinding halt?

Wouldn't you find it rather odd, if a group of people chose to set themselves apart as devoted followers of Darwin, even to the point of remembering his birthday, yet all the while thumbing their nose at the scientific paper he worked so hard to produce?

> By following my work, I would like you to learn how to distinguish between the term 'Darwinian evolution' and the term 'scientific evolution.' If you can master this one small thing, you will be well and truly on the way, to letting Charles Darwin **OFF THE HOOK.**

As I have done previously, I am now ready to move on to the next chapter, (to fill in the gaps). However I feel that the reader can benefit, at this stage, by taking a quick look ahead. To discover what important points are covered in the next few chapters...

In chapter two I will expand on the ideas of Darwin, in particular the all-important fossil record.

I present a simple question, what is missing from the fossil record *and* how should science continue if the expected [evidence] simply isn't there to be found? I set out to explain how and why it is so vital (for the theory) to find the examples that Darwin was seeking. It's a matter of simple deduction really; if the evidence isn't there then the [scientific] theory collapses.

Chapter three is the most important, showing the thoughts recorded by Darwin, as he considered the possibility of finding a symbiotic relationship (in present day life), as well as *how* such a discovery would affect his theory.

I also set out to explain why it would be so difficult (almost impossible) for Darwin to accept such a relationship. The problem (of relationships in nature) has a multiplying effect (on the calculations), making the outcome 'exponentially' more difficult to be accepted. This is done, simply by applying a scientific approach (to the information), in the way that the man–kind would expect.

In chapter four; I give a brief summary of the poor attitude towards Darwin's work, showing that the scientific establishment must *thumb their nose* at Darwin's work.

- Strike one: Scientific evolution opposes the Darwinian view of the human eye.
- Strike two: Scientific evolution opposes the Darwinian view of the fossil record.
- Strike three: Scientific evolution opposes the Darwinian statement concerning mutualism.

Chapter Five has the story of the history of the human foot, explaining how the big–toe might have "came–about" by a process of scientific evolution.

In chapter six I discuss a known [random] change to a particular product's DNA. This chapter will clearly show just how unlikely it would be to achieve an improvement by the slow and steady *change* of evolution.

In chapter seven you will learn that, for an existing creature to evolve into something else it must be supplied with new and novel information. It also shows that there is no known process on this planet that could provide the new and novel information.

In Chapter eight, I will expand on the idea of the development of the human. If anyone can grasp the basic idea of the changes needed—to produce a different type of being—they will then more easily understand what is missing from the fossil record.

In this first chapter the thoughts of a well–known 'crackpot' have been shown for all to see...

> Sylvia Baker was taught that she had to...follow the evidence wherever it may lead; EXCEPT WHEN THE EVIDENCE DID NOT SUPPORT THE CURRENT PARADIGM.

Chapter Two

THE FOSSIL RECORD

Charles Darwin:

According to this theory innumerable transitional forms must have existed. Why don't we find them embedded in countless numbers in the crust of the earth? It will be more convenient to discuss this question in the chapter on the imperfection of the geological record; and I will here only state that I believe the answer mainly lies in the record being incomparably less perfect than is generally supposed. The crust of the earth is a vast museum; but the natural collections have been imperfectly made, and only at long intervals of time.
[ORIGIN OF SPECIES, chapter 6,
Difficulties of the Theory, p161]

To truly appreciate the theory of evolution, each scientist must include a study of the fossil record. In fact whenever a spokesman for science is 'caught out' on evolution he will automatically fall back on the word 'Darwinian' and the all–important fossil record.

Darwin was honest about the fossil record he called it imperfect. He believed that the progress of scientific discovery would lead to finding thousands of intermediary 'products.' Vast amounts of imagination have been used to 'discover' (any) single example of a *genuine* half–baked intermediary. None have ever come close, not one!

Evolutionists desperately squabble over which 'finished product' might or might not be a candidate for a tiny fraction of the 'so-called' evidence. Now, well over a hundred years later and science is still eagerly anticipating a single *genuine* transitional form.

Rather than accepting the words of Darwin, science simply puts the transitional problem aside as if Darwin had never thought of it. However, the idea of abandoning the transitional forms is not acceptable to Darwinian evolution. Transitional forms **must** be found in great quantities, far outweighing the number of finished products.

Simply put, if the half–baked products remain elusive then the evidence for Darwinian evolution is (sadly) lacking. When it comes to the fossil record, it is time for science to admit defeat and move on.

Instead of morphing to some other form of Darwinism, science could gain some credibility by removing Darwin from the scientific–community and admit the truth to the general public, ending the deceptive practice of never ending 'Darwinian' explanations.

It would be very complex indeed to study the transitional stages of the human eye, so for something a little easier to study, I will (in a later chapter) include the study of the human foot. Human bones are a more obvious subject to study, and intermediate forms are very likely to leave a trail of specimens for science to 'discover,' not just one or two bones from a single 'survival,' but hundreds of bones from hundreds of steps along the way. If you are willing to contemplate the arrival of the human foot in an evolutionary way, you will be staggered at what was expected (when compared to what has been discovered).

The idea that would have haunted Darwin as he wrote "Origin of Species" would be the idea that the human foot evolved from a monkey's cousin. All creatures related to the monkey's cousin have a 'hand–like' foot with a thumb in place of a big toe. To cover how this change might have happened, I will include some simple math about how any change *could* occur over time. If you believe in mathematics you will clearly see, there would need to be a large number of unwanted and undesirable changes along the way. Where are the fossil bones to support this theory?

The fossil record, (keeping pace with evolution) would leave behind a massive number of failed experiments. If the experimental–foot wasn't suitable, it would still have bones, leaving a record of something unfit. Such a creature would have died shortly after jumping out of the trees. It would take several generations of dead bodies, (just down from the trees), before a bone structure for 'running' was accidentally achieved. It simply doesn't make any sense for the fossil record to contain only finished–products; such an outcome is illogical and unreasonable.

Science can offer no explanation for the missing records all they can say is that there is more work to be done.

Darwin himself didn't just shrug off the imperfection of the fossil record. He recorded his thoughts on paper. I can say with great confidence that if Charles Darwin were alive today he would be staggered by the lack of evidence.

The very evidence that he was anticipating is completely missing. Not one of the transitional forms has been left in the record for 'science' to discover.

Australopithecus is an extinct genus of African hominid. When scientists discovered this hominid they produced an upright model of the creature.

Science was so desperate to find an intermediary that they made the model of the hominid with a human–like foot. It has since been proved beyond any doubt that the hominid bone (definitely part of the foot and definitely part of a finished product) could not possibly support a forward pointing big–toe. To the shame of science, the model of Australopithecus was left on display with a human–like foot for some time, even after the gross error was discovered.

Yet putting a completed (human–like) foot on a false discovery would NOT make an intermediary specimen in any case! It still wouldn't fit with 'gradual change over vast periods of time,' as the science goes. If the big toe moved slowly from one to another it would leave a trail of evidence. If the big toe moved randomly from one to another it would leave some horribly disfigured feet along the way, the failures are just as likely to leave bones as the successes, so where are these failures?

Science should not be desperately clutching at straws to find an intermediate–form; they should be discovering such things by the hundreds. It is fair and honest to say that without any intermediate–form, science must accept that for evolution to be true, the monkey's cousin must have produced a human like foot in just one attempt. Producing a new and improved 'finished product' in just one attempt does not fit in with the scientific work of Charles Darwin.

A new product developed from a completely different concept all the way to a working model, in just one attempt! Does that sound anything like a gradual process over vast periods of time? NO IT DOESN'T.

If you choose to—you can *see* what's missing. Once you understand *why* there would need to be hundreds of examples (of failed changes), the problem of missing bones (in the fossil record) will become something real for you to consider.

Survival of the fittest implies the non-survival of the things that were unfit. Survival of the one–and–only product in just one–go cannot be passed on as Darwinian evolution. The scientific establishment would find evolution much easier to cover if they hadn't used Darwin as a scapegoat.

The system of giving evolution the big green tick of approval does not fit well with the *SCIENTIFIC METHOD*. Science must someday move–on, to teach science as it is observed in the lab, when that happens the label 'Darwinist' will fade away forever.

A famous paleontologist – Dr. Gaylord Simpson freely stated, "...absence of transitional forms is not confined to mammals, but is an almost universal phenomenon, as has long been noted by paleontologists. It is true of almost all orders of all classes of animals..."

[Tempo and Mode in Evolution, p. 105]

A few years later – Dr. Simpson was forced to admit, "It is thus possible to claim that such transitions are not recorded because they did not exist"

[The Meaning of Evolution, p. 231]

The sad lack of evidence eventually led to a little 'tweaking' of the theory. This was done with the development of a **new phrase**: Punctuated equilibrium. Wait on—is it slow and steady like evolution (gradual change over vast era)—or is it punctuated, like when you look back in the fossil record to the first appearance of the dinosaurs?

I guess science claims whichever one fits the argument at any given time. When the dinosaurs first appeared on the scene, (Triassic) there were about 20 different models of these creatures, they all seemed to pop up around the same time and they didn't show any signs of being linked from one era to the next. That's what I call punctuated!

So evolution is truly adaptable, when science wants it to be slow and steady, over vast time, then by golly it's slow and steady.

And when science **doesn't** want evolution to be slow and steady it automatically becomes punctuated, isn't evolution wonderful?

Chapter Three

SYMBIOTIC EARTH

Natural selection cannot possibly produce any modification in a species exclusively for the good of another species; though throughout nature one species incessantly takes advantage of, and profits by the structures of others. But natural selection can and does often produce structures for the direct injury of other animals, as we see in the fang of the adder, and in the ovipositor of the ichneumon, by which its eggs are deposited in the living bodies of other insects.

If it could be proved that any part of the structure of any one species had been formed for the exclusive good of another species, it would annihilate my theory, for such could not have been produced through natural selection.

[_Origin of Species, Chapter 6 - Difficulties of the Theory, page 193_]

Well, I feel that I have the right to ask, have you ever been taught **that theory** in any science lesson?

If you are a reasonable person, you should be able to admit (to yourself) the following. If Charles Darwin was alive today— and if he was accepted briefly by the scientific– community, he would rightfully teach that **natural selection could not possibly produce any modification for the exclusive good of a different species.**

Charles Darwin would teach that the **Darwinian theory of evolution has been annihilated.** So, how long do you think Darwin would last as a science–teacher? How long do you think he would last in a university, learning all about Darwinian evolution? Poor old Charles Robert Darwin would be classified as a crackpot in no time at all.

That's right Charles Darwin would find no place in science either as a student or as a teacher! Charles Darwin would insist on telling the truth about **his** theory!

Symbiosis:

The human body has a reliance on 'good' bacteria for their digestion and well–being. You've seen the ad on television where the bad bacteria outweigh the good! Humans today can take a capsule containing millions of cultured bacteria. These little fellows are not directly related to humans, they 'evolved' separately.

The bacteria that I am talking about did not form inside our body they came from an outside source, yet we can swallow that capsule with great confidence. The bacteria, once taken rely on us for their survival. Humans also rely on the bacteria for their own survival. This is a symbiotic relationship. This is also a life and death relationship. Without this form of symbiosis our lives would be cut short by illness and infection.

Another life–and–death relationship for humans is the story of our protein intake. In a later chapter I will briefly cover the story of amino acids. It is important to remember that there are 9 amino acids that our human body cannot manufacture. These can only be obtained from high protein food (such as beef or lamb or other high protein foods like poultry, fish or dairy).

Think of the information above from an evolutionary point of view. According to scientific evolution the *food* and the *human* evolved from a common ancestor via a series of random changes, leading to a survival of the fittest.

Do you think that cattle could have produced the spare parts for our bodies by a process of gradual random changes over long periods of time? The most logical outcome is that the random parts would be unsuitable. How on earth could the beef or lamb have produced 9 useful spare parts **for our benefit**, even if they wanted to? Random–changes means random–changes, the chance of these exact spare parts being manufactured by accident is effectively zero.

Humans needed these parts right from the get–go, without these 9 amino acids in the human diet a person would only live for a few years! If you look at the parts of the world where protein is not readily available, you will find that little children are dying before school–age.

It may seem as though I am going over the same point, I am merely trying to get you to 'see' the same problem that I can clearly see. So let's take a look at two important alternatives...

The first alternative: the grass and the cattle.

By definition the first living cell was much simpler (in its construction) than any cells that are observable today. By that reasoning the 'teamwork' of the grass growing food, and the cattle converting the food, will produce new and novel parts which (quite by chance) are the parts needed for us to stay alive?

Science has never shown a model that would produce these parts, because the math for such a model tends to look a lot like zero–chance. But if you choose this method then this near impossible process, from simple to complex must have been extremely easy. Therefore evolution is easy!

The second alternative: the parts were already made.

This model requires a first living cell (or one of its early cousins) to be horrendously complex. When you try to model a cell with all the spare parts (amino acids) completed and in the correct order (right from the get–go) the almost impossible chance flips over to **definitely impossible.**

In any case the process would definitely need a method that produced the parts required, and it had to be incredibly easy for random changes to achieve that sort of outcome.

Now that there is some known information to work with, take a look at an alternative. Of course for a scientist, there is no alternative, but what about you? As a fellow member of the human race you are entitled to consider data from a scientific viewpoint, **regardless of what you are told by the scientific establishment.**

This is all about how easy something can occur and how easy it is to understand (and believe). The main theme for the atheist is that he must NOT believe in a being which exists in another dimension. That's the number one rule of evolutionary science!

We know that science is allowed to use 2D shapes for complex calculations. You must also know that no–one has ever seen a 2D object! When you draw a triangle, the paper that you draw on has thickness, the ink or pencil on the page has thickness. The thickness (of the chunks of pencil on the page), is 3D. Humans can only see things that are 3D.

So even though the 2D objects cannot be seen they are very useful in our everyday lives. The 3rd dimension is the world that we *can* see. The 4th dimension is the space–time continuum, but what about the 5th dimension? Should we dismiss the 5th dimension just because we cannot observe it?

Science is willing to accept that there are other dimensions— but they have certified (beyond reasonable doubt) that there are no living beings in that 'place' they cannot see.

What if we do believe that there is another dimension? That's acceptable to science, but if there is something alive in the 5th dimension – then *IT* defies the laws of science.

On what grounds do they hold that belief? Science tells us they are 'sure–and–certain' that evolution can produce a first living cell here on earth. They assure us that evolution is easy – (it has to be)! But if you look to another dimension where matter is different, where there is much more available energy, then evolution just cannot work at all, ever.

Can you understand how that law works? To me it appears as though the people of the scientific establishment are manipulating their own thoughts to achieve a predetermined outcome. Not a scientific outcome.

A man–made outcome!

For OUR logical mind, if evolution became staggeringly easy on a lonely planet in the third dimension, then it would also be easy in a 5th or 6th dimension (with a different kind of time and matter and an abundance of energy). Science has offered no explanation for this segregation of thought. Science should be a one size fits all approach, but that's taking things a bit too far.

The hardened atheist scientist MUST believe that evolution is incredibly logical and incredible easy. Yet if they allow their mind to drift off, if they allow themselves to ponder the 5th dimension, they must now believe that evolution has been impossible since before the beginning of time! Now you know why I often say, "Their mind must squirm like spaghetti." The mind should never be put through that kind of torment.

Imagine if you had to spend all day convincing your–self that evolution is a fact. Spend all day believing that the process required for evolution happens by chance, that the process happens very easily. Then at the end of the day, some clown comes along and mentions the 5th dimension. Now you have to suppress these thoughts. Now (with your thoughts in another dimension) evolution suddenly becomes, not just difficult, but completely impossible. You have to *know* that it just couldn't have happened!

You will also notice that I keep claiming that the evolutionist is a double–minded character. The information above is a fine example of double standards. The complexity of scientific evolution is being manipulated to fit the need. One minute, evolution is incredibly easy and natural. The next, it is incredibly difficult, even impossible.

Whenever the 'evolution' needs to be ridiculously easy, it is. Whenever the evolution needs to be impossible then, you can bet it's impossible. Isn't evolution wonderful?

Observable Symbiosis:

A fine example of symbiosis (found in PNG) is the Boxer crab. It carries a pair of small anemones in its claws, when approached by a predator the crab moves the stinging tentacles of the anemones to deter the intruder. The crab has a clear benefit provided by the anemones. The anemones benefit by consuming particles of food dropped by the crab at feeding time. These are obviously reliant on each other.

A relationship that many divers have encountered is cleaning–symbiosis, a widespread form of mutualism, common to temperate and tropical waters. Gobies and shrimps are fish-cleaning species. The fish being cleaned are called client–fish and the sites they go to are known as 'cleaning stations...' In tropical waters these areas of the reef are clearly defined. Divers see a number of stationary fish and their attendant cleaner organisms together in one area.

So why is the mutual relationship of the crab and the anemone so concerning for the student of 'Darwinian evolution?' (Apart from the genuine statement made by Darwin himself, **that it would annihilate the theory**).

The reason for the concern with symbiosis is this: For one creature to reach the point of being a "finished product," [by a process of evolution] is "extremely unlikely." But whenever you find two creatures, which became "finished products" at the same time, (and are interdependent in some way) then you have a new formula... The calculation (for the chance of both succeeding) now becomes "extremely unlikely" **multiplied by** "extremely unlikely!"

I believe that Charles Darwin was correct when he recorded these concerns about symbiosis, but for the scientific establishment such a claim is counted as nonsense.

The most important point I can make here is that the scientific establishment is dancing up and down about Darwinian evolution (even to the point of recalling his 200th birthday) while actually thumbing their nose at Darwinian 'science.' What a motley crew!

I say (with all my might), to the scientific establishment, either remove Darwin's name from your 'scientific' theory OR start teaching the science of Charles Robert Darwin. You must stop calling out Darwin's name whenever you encounter a difficult situation, it's time for you to stop being a "Darwinist" and return to science.

Of course the above is just a rant. I don't actually want any scientist to read my work—because they will trample on my ideas as quickly as they would trample on Charles Darwin himself. The idea of any scientist or any science-worshipper reading this book makes my hair stand on end!

Freeloading Caterpillars Masquerade as Ants:

There's something about the caterpillars of some large blue butterflies that certain ants find irresistible. A deceptive chemical odor apparently convinces the ants that the caterpillars are long lost kin. The ants take them home, and the adopted caterpillars then make a meal of the ants' own young. Tumbling from plants into the foraging area of their main hosts (*Myrmica ants*), the caterpillars secrete the same recognition-chemicals as the ant larvae, according to Jeremy Thomas of Britain's Institute of Terrestrial Ecology. The ants soon discover the caterpillars and haul them back to their nest.

Once inside, the party starts. One **primitive** caterpillar species simply devours the ant larvae. But a second **more advanced** type rears up like an ant grub, to be fed by workers. Large blue butterflies, all endangered by habitat–loss, can use such an advantage.

[National Geographic, December 1991]

The scientific theory of evolution relies on slow random changes over vast periods of time. If your logic circuits are alive and well, you will easily see the problem with this ant and caterpillar story. One species developing a scent that was distinct and very useful is almost–impossible to achieve. Two species developing the same function in the same time–frame can only be represented as almost–impossible, **multiplied by**, almost–impossible. The chance of such a thing actually occurring in nature is effectively zero.

If you can find an old copy of Australian Geographic for OCT–DEC 1991, you will find the heading 'Sweet Liaison.' The story of the sugar ant and the blue butterfly includes 16 pages of text and illustrations. This Australian story is a special example of the difficulties encountered when science tries to shoe–horn *all* information into the realm of scientific evolution.

You will find many hundreds of symbiotic relationships on this planet. In fact the majority of scientists promote the idea of an **ecosystem**. The theory of scientific evolution requires a survival of the fittest mentality. An ecosystem implies co-dependency, and co–operation among its inhabitants. These two themes are not compatible!

At this point, I can assure you that I could produce chapter after chapter of information from the work of Charles Darwin, and each time there would be a staggering difference between Darwinian thought and scientific thought. If you cannot let Darwin off–the–hook, after these three important points then another twenty chapters would make no difference whatsoever!

Richard Dawkins is one of the main champions of the theory of evolution. Whenever Dawkins is 'caught out' by a difficult question (about the impossible nature of evolution), he always refers to the expert – Charles Darwin. That reference is no longer valid, as the 'Darwinian' theory of evolution has now been 'officially' **annihilated.**

For the scientist and the peddler of all things scientific, Charles Darwin is the hero of the day. But for dumb stupid humans of the mankind – Darwinian Theory is forbidden. If I start talking about the work of Charles Darwin as though it had scientific merit, I am automatically crackpotised. If a scientist gives credit to Darwin's work he is talking about a different Darwin, he is talking about **another Darwin who was invented by science** for the purpose of deception.

Chapter Four

THREE STRIKES

<u>Scientific Evolution:</u>

In Chapter one, the vast difference in the two theories is exposed by looking at the scientific evidence for the evolution of the human eye. Perhaps it would be more accurate to say–the LACK OF EVIDENCE.

Try to imagine vast rooms inside the scientific–establishment; the first room contains all the scientific evidence gained by the study of a finished product, the human eye. The sign on the door is "**THE HUMAN EYE.**" Under the heading on the entry door there is a warning for future students…

This room is dedicated to the study of the human eye. Charles Darwin says "To suppose that the eye with all its inimitable contrivances for adjusting the focus to different distances, for admitting different amounts of light, and for the correction of spherical and chromatic aberration, could have been formed by natural selection, seems, I freely confess, **absurd in the highest degree**."

For the rest of the sign to be "scientific" the following statement is made...

The human eye evolved, evolution is a fact, if you have a problem with Darwin's theory just get over it and move–on.

Then in Chapter two, the vast difference in theories is exposed in another way, by looking at the (lack of) scientific evidence found in the fossil record.

Moving along to the second room, there is another heading and another warning to future students...

The sign on the door is "**THE FOSSIL RECORD**," (because the room is dedicated to the study of fossils). Under the heading on the entry door there is a warning for future students...

Charles Darwin cries out for physical evidence to confirm the existence of thousands of intermediary species. If the in–between creatures cannot be found it would bring **an abrupt end to Darwinian evolution**.

For the rest of the sign to be "scientific" the following statement is made...

The fossil record obviously contains a list of finished products. The fossil record ALSO contains bits and pieces originating from finished creatures, but the fossil record does not contain any half–baked creature of any kind. In spite of the lack of evidence you must believe that, "the various fossils on display were formed by a process of scientific–evolution." If you have a problem with the theory, just get over it and move on.

In Chapter three, the vast difference in the two theories is exposed, by looking at the scientific study of symbiosis, in particular the theory of mutualism.

Moving along to the third room we find that the room is completely empty (and soundly locked up forever). On the door is the title **"SYMBIOSIS."** Under the heading on the locked door there is a warning for future students...

Our reason for concern with the theory of mutualism is this: For one creature to reach the point of being a finished–product, is "extremely unlikely." But whenever you find two creatures becoming "finished products" at the same time, (and are interdependent) then you have a new formula... The calculation for **chance of success now becomes extremely unlikely, multiplied by, extremely unlikely!** Charles Darwin was right when he wrote these truthful concerns about symbiosis.

For the rest of the sign to be "scientific" the following statement is made...

Science has determined that the theory of mutualism is unsound. This room will always remain empty to celebrate this scientific breakthrough. Symbiosis must always remain an unproven theory. This room must never contain evidence for mutualism. This room will always remain locked to ensure that no future discovery will be allowed to disrupt **our** scientific theory.

As I said in chapter three, I could go on and on about the differences between Darwinian evolution and scientific evolution. I am not going to do that. If you cannot see the vast differences that have been exposed in the first three chapters then all the books in the world will not make a dent in your hardened attitude. **Reading further into the details of this book would be utterly pointless for any person who believes that Darwinian evolution has a genuine home in the scientific establishment.**

For the remainder of this book I will mostly be referring to "scientific evolution" If the word "Darwin" appears in any material please cast your mind back to the early matter of this book. I believe that Charles Darwin is a real scientist and is well and truly **off-the-hook**.

For example I might use the words "In Darwin's time," and then follow on with a comparison to modern science. In that case I am not referring to an error (on Darwin's part), **I am** referring to an era, (a period of time). The following is an example of the difference between the science of Darwin's time and the science of "our time..."

The text written out below is quoted from a video that is available to watch online. The video has this title...

PREMISE MEDIA presents
A Rampant Films Production

"NO INTELLIGENCE ALLOWED"
Starring Ben Stein

About a third of the way in, there is a very brief cartoon about the 'success' of the first cell, then...

"We don't know what caused life to arise, did it arise from a purely undirected process? Or did it arise from some kind of intelligent guidance or design? The rules of science are being applied to actually foreclose one of the two possible answers to that very fundamental, and basic, and important question."

Ben Stein; "So the rules of science say... we will consider any possibility except one that is guided?"

"Exactly"

Ben Stein; "No matter how life began – on the backs of crystals, or in the test tube of some intelligent designer, everyone agrees it started with a single cell. But what is a cell?"

Next interview; "Let me ask you a question…

Darwin wrote 'The Origin of Species' – published in 1859. He had an idea of the cell as being quite simple, correct?

"Yeah everybody did."

"If he thought of the cell as being a Buick, what is the cell now in terms of its complexity by comparison?"

"A galaxy"

Next interview; "If Darwin thought a cell was, say a mud hut, what do we now know a cell is?"

"More complicated than… a Saturn Five"

If you do watch the video, don't just stop there. The section shown (above) is followed by a most amazing video, including an animation of a string of amino acids entering a chamber where it is folded into a protein molecule. This is an amazing thing to watch, even for a hardened scientist.

So Darwin is not the culprit here, he is just the timestamp. This chapter 'wraps–up' the first three chapters by using the old saying… "Three strikes and you're out."

- Strike one: Scientific evolution opposes the Darwinian view of the human eye.
- Strike two: Scientific evolution opposes the Darwinian view of the fossil record.

- Strike three: Scientific evolution opposes the Darwinian view – by identifying mutualism.

Darwin specifically wrote, "If it could be proved that any part of the structure of any one species had been formed for the exclusive good of another species, it would annihilate my theory, for such could not have been produced through natural selection.

[Origin of Species, Chapter 6 - Difficulties of the Theory, page 193].

There are many forms of "Darwinism." There are many forms of "Neo–Darwinism," even the hardened evolutionist has trouble deciding on which version is the one they follow. For Darwin's work to be scientific it must not be a moving target, it must not be a one–size–fits–all kind of exercise. Truth doesn't change from day to day. One can only uphold a belief in the 'flat earth' theory until one discovers revolving celestial bodies. One can only uphold their belief in Darwinism if they learn what Darwin actually taught, and then support the **truth** of his work.

Darwin himself pointed out several difficulties with his theory. Science owes it to Darwin to respect those difficulties – right up to the point where those difficulties are fully explained (in a truly scientific way). This has never been done!

The scientist who upholds his belief in Darwinism has a credibility gap. He doesn't have a credibility gap with his fellow scientists – because they are [all] playing the same tune. He doesn't have a credibility gap with the lesser human beings (those of the man–kind) because he dictates to them and doesn't have to listen to them anyway!

No, he has a credibility gap with himself.

He has to travel to work each morning fully convinced that he is a genuine Darwinian scientist, presenting Darwin–himself as the master of evolution. But when he discovers that Darwin was concerned about how the eye might have evolved, or when he discovers that Darwin was looking for a thousand intermediate steps – (clearly defined in the fossil record), or even when he discovers an example of mutualism, his thoughts become compromised. He must ask himself "which Darwin do I follow?"

"Do I follow the scientific version of Darwin or do I follow the 'Charles' version of Darwin?" In the end the poor old scientist has to compromise, he has to become double–minded, because he has to follow BOTH OF THEM.

A scientist at any level should never be put in a position of such compromise.

Chapter Five

ONE IS NEVER ENOUGH

What about your own belief system? If you want to follow Darwin, if you want to be a Darwinist, or any sort of Darwinian anything, then you must insist that 'Darwinian theory' has been annihilated for all time.

The scientific theory of evolution cries out for a scientific explanation. In this chapter we don't just *say* there is evidence missing from the fossil record, we actually take a look at what science *naturally* expects to find.

It is one thing to say that there is something missing from the science of the fossil record; but it becomes very real if the reader can 'see' what's missing and why such intermediate steps are so vitally important (for scientific evolution to be true). There should be no desperate attempts to contrive explanations, like the ones we have seen with the story of Australopithecus.

For those who are unfamiliar with the nonsense that was blurted out (by elitists who should have known better), I have included here a refreshing story of blunder upon blunder by the scientific community. The first blunder was the act of showing the world just how desperate the scientific community was, to find an intermediate creature. The second blunder was to proceed with **something that *wasn't*** as though it was **something that *was*.**

For those of us still in the man–kind, selling a false idea as if it were true would be counted as an outright act of deception. When any elitist group willfully sets out to deceive the masses they have overstepped the boundary of science.

On the other hand, I am not setting out to deceive you with technical explanations. All I hope to achieve is for you to *see* the scientific–spin that MUST be used to fabricate a scientific theory.

Remember that there are two parts to this evidence (or lack of evidence). The first part looks at the willful deception used by those studying the fossil record. The second part, which is even more staggering, looks at the what–if of the whole sordid affair. I ask the question, what if science had been correct in their wild and desperate grab at the Australopithecus fossils, (for the purpose of supporting the science of evolution). I hope that you will easily see that success in finding this thing (as an intermediary) would have been an even greater deception.

If Australopithecus were a genuine human–like creature would that help their cause, or make things worse?

Now back to the fiasco!

What does a desperate attempt look like; and why is the human mind so easily fooled?

Well, I can answer the first question clearly and accurately, the second part I will leave for the scientist to explain.

The problem began when science discovered some bones that didn't fit well into a particular category (in the fossil record). The unexplained evidence suddenly became 'evidence for evolution.' Now you might think that science and magic just go naturally together, but I was taught that science must avoid magic and present facts. Yet, in the case of the fossil record, facts have been tossed aside and replaced with magical assumptions.

The thing that makes this so shameful is that the magical assumptions were required ONLY because the scientific theory of evolution was involved. Take evolution out of the picture and the desperate assumptions would NOT be required. Without evolution at the helm, science would have 'seen' the evidence that they had gathered, and then the results would flow–on **from the evidence itself.**

Fitting the evidence into a foregone conclusion is hardly the act of a scientific mind. If an old man living outside of the walls of science can see that the evidence is false then surely someone with a scientific mind can see the same thing from the same evidence. **So what exactly is missing?**

Every finished product, in fact every single part of every finished product (in the fossil record) required a time before. So let's take a (**mathematical**) look at what the man–of–science would expect to find in the evolutionary tree of life. Knowing that if such evidence had been found it would have been shouted from the rooftops.

Muttering about the lack of evidence is not a scientific solution. The most common comeback from science is that the missing information is just a 'research problem.' Baloney; There are hundreds of steps required, and only **finished** specimens have been discovered? Notice the word finished. What's going on here? Is it mass hypnosis?

> This information could well be included in chapter eight, where I take a scientific look at one of the suggestions made by Richard Dawkins. (The matter was briefly mentioned in the introductory book, also in chapter eight). When you get to chapter eight, think back on this section and remember the mathematics. Mathematics is real; Mathematics doesn't lie.

> It's not me calling for mathematical–explanations; it is the nature of the theory of evolution.

Evolution DEMANDS a natural explanation. I hope you have some understanding of the science of micro and macro, as Richard believes that "lots of little bits of 'micro' over very, very long periods of time, add up to a 'macro–advance' in any particular creature.

For example, let me quote from one of the books by Richard Dawkins, who is a famous atheist and evolutionist:

"Well, I must mention the alleged distinction between macroevolution and microevolution. I say 'alleged' because my own view is that macroevolution (evolution on the grand scale of millions of years) is simply what you get when microevolution (evolution on the scale of individual lifetimes) is allowed to go on for millions of years... I have never seen any good reason to doubt the following proposition:"

"MACROEVELUTION is lots of little bits of MICROEVOLUTION joined end to end, over geological time, and *detected by FOSSILS* instead of genetic sampling."

[The Ancestor's Tale - A Pilgrimage to the Dawn of Evolution, pages 603 and 605]

If any scientist allowed himself the luxury of studying the fossils, WITHOUT THE FOREGONE CONCLUSIONS OF EVOLUTION, this is how he might proceed.

Studying the evolution of the eye from one type to another would be difficult indeed, and one wouldn't expect to find much evidence in any case. But what about a (simple) structure such as the big–toe of a human?

Now science has something specific to study, now it has something that MUST have happened over long periods of time (for evolution to be a truthful account). It is also something that would leave a specific trail of evidence, the very stuff that science lives and breathes, or at least it should.

The precursor for the human big–toe is a thumb–toe suitable for climbing and hanging in trees. The famous Lucy was displayed in a museum with human–like feet. The bones themselves were not human–like at all, they were ape–like. For a period of 20 years science has gone back and forth with the story of an Australopithecus named Lucy. So for the sake of this exercise I am going to 'give' them LUCY as a finished product (with a human–like foot), even though that assumption is NOT supported by the science of the fossil record. (The human–like bone was found at a different location and assumed to be the evidence they needed.

So giving *them* a finished–product with human–like feet, DOES NOT HELP THEIR CAUSE, all it does is present us with another finished product that came about in one go. Now let's look at what *WOULD* be required, to satisfy the scientific theory of evolution.

Have you got it yet? The slow progression from a thumb–like toe to a human toe would definitely leave a trail of hundreds of failures along the way. We can't talk about the survival of the fittest without studying the pre–Lucy animals as they morphed from one type of DNA to another over a very, very long period of time.

So finally this is what science is looking for (in the fossil record). No, not just another finished product like us, but a trial and error (random change) progression of information that provides evidence for GRADUAL change. Not just one random intermediate that can be made to fit their mind, but **real bones from real failures**.

If you have your shoes off, place your hand over your foot in such a way that four fingers line up with the ends of four toes. Now move your thumb around to 90 degrees and take a look at the difference between a thumb–like toe and a forward pointing toe. Now slowly count to nine while moving the thumb to the same line as the big toe. We'll call each count 10 steps towards a new footprint (90 steps altogether). The first thing you will notice is that the thumb is now twisted around to one side (almost 45 degrees). Now without changing the fingers much, try to bring the thumb flat and forward pointing, **just like your big toe.**

Doing these changes by random mutation would NOT restrict the changes to these 90 mentioned or these 45 mentioned, because random is random. For the benefit of scientific evolution we will allow the assumption that somehow the monkey's cousin knew what it was aiming for, thus allowing us to look only for fossils that fit the foregone conclusion.

Using the above method of random and gradual change over vast periods of time we can calculate how many fossils will be found (in the fossil record).

To produce a human child with a big toe you first produce two parents, each one with a pattern for a big toe, so how many potential parents are required altogether?

Now we have the values we need to set out a plan for the kind of evidence we would definitely expect to find (in the fossil record). The number of failed creatures on the male side is 90 times 45. The number of failed creatures on the female side is 90 times 45. So with gradual random change, over time, we should expect to find over 4000 specimens of males and over 4000 specimens of females. Almost all of these would have died a slow and painful death as soon as they come down out of the trees. Each one born would have denied the original parents a natural child (of their own species). Each mother would have been horrified at the defective child she had produced. For this to occur on such a grand scale would have led to an extinction event. The existing species would lose out because they are no longer producing children true to their existing DNA. They would simply have no—one to take *their* place.

As well as the above problem the first of the new species (the 1 in 4000 success) would be a lone survivor without a partner. There would need to be thousands of attempts (all over again) with the hope of achieving another success, this time of the opposite sex!

Only the one that morphed with a strong forward–pointing big–toe would be able to run along the ground. The other 4000 (missing) fossils should be very easy to find. The expectation is that they would have been killed almost as soon as they arrived, or very soon after. The new half baked specimens could no longer hang in trees and they could not expect to run along the ground. The shape and position of the big toe is the key to their success.

So what must be produced in the fossil record? Scientists would expect to find a very large pile of dead bodies (for the correct male to be produced by random change). Plus they must expect a very large pile of dead bodies (for the correct female to be produced by random change). We know this for sure—the one with the thumb like toe (still at 90 degrees) would be unable to run. It would be unable to track and catch food, unable to escape a predator, and so on.

Altogether over 8000 failed attempts to get just one pair with a human–like foot. Then you must also consider the following. What are the chances for the two improved specimens being pumped out in the same location and in the same generation?

You already know what I think about that. Any reasonable person could not possibly believe that all those thousands of random changes could occur (in the real world) without leaving any evidence whatsoever. The one finished product that science is crowing about is totally inadequate because it appears in just one step.

If you can bring yourself to believe that such a change happened in just one step then you are capable of believing anything. If you still believe in random change over time (after considering the fossil record) then you have a scientific mind that cannot be changed. You are wasting your time reading this book because you are not allowing your mind to process the truth.

On this planet the worst kind of scientist is one with a credibility gap. A scientist with a credibility gap is capable of lying to himself (about the fossil record) and then believing his own lie? Anyone who is happy to dwell in the man–kind should not have such a credibility gap. I have no reason to fool myself about evolution I simply follow the strong scientific evidence (or lack thereof).

The thing that should be most outstanding in your mind is the time period involved. If we give science the facts about Lucy's foot exactly as they suggest then we have a double standard. Science wants this example to fill in the gaps for human evolution. But is that achieved? No, the exact opposite is achieved. Even this 'good' evidence exposes the fact that the gaps are much worse than expected.

Making Lucy's foot a human–like foot, is a one-jump wonder. Only a person with their head in the clouds could believe otherwise. Bring yourself down out of the cloud of science and see the evidence as it is presented, rather than absorbing the scientific spin.

What is presented by science, is a monkey's–cousin with an unchanged foot (and in the same era) a monkey's–cousin with a human–like foot.

The basic pattern for the ape–like foot has remained unchanged for over three million years. The basic pattern for the human–like foot has remained unchanged for over three million years. How can anyone hope to rationalize that evidence (the instant new foot pattern) as slow and gradual change over vast periods of time?

There is nothing in the DNA of modern apes (or chimps) that suggests they are capable of experimenting with different foot patterns. There is ZERO evidence to support gradual change, either before during or after the appearance of a new and novel "completed" Lucy-foot. The new foot–pattern is coded in very accurate (modulo four) arithmetic and **the scientific community has nothing to offer which could explain how the new pattern came into existence.**

Chapter Six

SCIENTIFIC EVOLUTION

Now you are set free from the term 'Darwinian.' Along the way, you may have noticed the slippery nature of the work going on inside the culture of evolution.

Our interest in the origin of life is very important to us, we should investigate we should inquire. What we must **NOT** do is just blindly accept whatever we are told.

Many of the things that you have learned about evolution are based on assumption. Science assumes that someone has given evolution **the big green tick of approval**. Many fields of science then proceed to gain knowledge, based on the 'evidence' supporting the scientific theory.

In science that should not be happening, each field of science must be able to discover evidence and then report on the evidence (based on the discovery), rather than based on some "ism" of some kind, like Darwinism for example.

For an imaginary world (without evolution at the helm) the discovery of human DNA (in 1953), would have been a staggering event. Scientists would have shouted from the rooftops that modulo–four coding was around, no doubt millions of years ago. Many millions of years before humans invented binary digits.

The two–bit coding (invented by humans) had to be on–hand before anyone could dare to imagine a digital–computer. Yet (in 1953) here was this most tightly packed chunk of information that science had ever encountered.

Yes, you would expect the scientific community to follow the '**information**' wherever it may lead; But there was one little problem—the existing theory of evolution.

No one (in any field of science) wanted to make a great deal of the coding that was discovered, why is that? Because coding sounds like design and design assumes a designer. Simply having a process that may have needed a designer would break the rules of scientific evolution.

Most people assume evolution to be true. They don't like anyone saying things that don't 'fit the mind.'

I'm hoping that at least some readers will gain something worthwhile from this work. How will you go, when you read things that don't fit with 'evolution,' things that you were taught to be true? Will you dismiss the facts without a second thought?

Some readers will be able to poke and prod at their world–view, while most will simply hold onto their beliefs. Either way it's your choice.

The first thing to discover, about the–science–of–evolution is the 'smoke and mirrors' way that a person is taken in. When a young person goes–on to study evolution, one of the 'tricks of the trade' is to demonstrate a mysterious thing called 'evolution–in–action' this is done by performing a repeatable experiment.

That all sounds very scientific, because it is repeatable. The experiment actually demonstrates 'bacterial resistance.'

The observer leaves that experiment, fully–convinced that they have seen evolution–in–action. However, if there were an expert present in the room, he would have explained what was actually witnessed. In fact the young subjects had just witnessed a programmed response.

What really occurred was the 'loss of' or the 'change–of' information, no new genetic material was seen by anybody, nothing had evolved, simply a dying creature 'automatically' trying to fend off a 'problem.' The idea of any scientist calling it 'evolution–in–action' is absurd in the highest degree. If you doubt that then try to consider this… Science can now study how this automatic reaction is controlled (by an internal mechanism).

Colin Patterson, (once a well–known fossil expert), had this to say… "One morning I woke up and something had happened in the night, and it struck me that I had been working on this stuff for twenty years, and there was not one thing I knew about it." He addressed his concerns to both the geology staff at the Field Museum of Natural History and the Evolutionary Morphology Seminar at the University of Chicago. "Can you tell me anything you know about evolution, any one thing that is true?" Each time, he was met with near silence. Someone at The Evolutionary Morphology Seminar, stated: "I do know one thing—it ought not to be taught in school." Patterson then said… "It does seem that the level of knowledge about evolution is remarkably shallow. We know it ought–not be taught in school, and that's about all we know."

All organisms have some sort of immune system, some means of protecting themselves from agents in the environment. Immune systems are extremely important for survival.

Innate systems rely on having information about the 'problem' before it arises. Adaptive immunity is stimulated by exposure to pathogens. The cell is ready to act by producing a form of protein known as antibodies, these attach to a pathogen leading to its elimination. The antigen receptors that are a part of this process are in place **before** the pathogen is encountered.

The changes that occur are described as blind and random, (by the scientific establishment). A process that has parts in place, in anticipation of an event – **cannot be described as blind or random.**

By recognising these tricks of the trade you are taking an important step forward, toward 'seeing' what is required for a faithful belief in evolution. A process that is obviously preset to occur, without causing any damage to the genome, is being presented as a random process. The changes that occur for the immune–response are specific and highly regulated. They are presented to the observer as random changes—such a thing can only be described as willful deception.

At this point I need to introduce you to some of the 'logical tricks' employed by the scientific establishment. The story about the bacteria is one little trick, but how does the *logic* work? (Logic must not become something that can be tweaked). Here are some potential ways that logic *could* be manipulated...

- o Misuse of Authority: This has already been covered in the first few chapters. A fine example would be when someone uses the NAME Darwin as a safety net for his personal belief in evolution, (rather than Darwin's work).

- o Appealing to the public: People try to sell science as a popular product. Science must never be done this way. Always deal with the facts. Humans *are* able to discern **from the facts**. The scientist must never claim they have the facts by claiming that they have the majority–support, these are two different things.

- o Begging the question: This happens when an assumption is used to validate a premise. This could be referred to as circular–reasoning.

- o Finding support in the future: "It's just a research problem." Science might claim that the reason there is no half–baked fossil–remains is that none have been found, YET? In reality Darwinian Science predicts hundreds of failures **for each step of the process**, each time leaving the fittest to survive! The story of the development of the human foot demonstrates **WHY** you should expect many hundreds of obvious failures.

o Generalization: A small amount of information is used to (prove) a much larger conclusion. A generalization would be if you landed in a city where every house had a motorbike in the driveway (instead of a car), then you come to the conclusion that all homes (in all cities) will have a motorbike out front?

o Hypothesis which doesn't fit the evidence: This is very important, because a scientific hypothesis that survives testing would eventually become a scientific–theory.

So what exactly are we talking about when we use the term evolution? Understanding the word evolution (and two or three of its meanings) is a basic foundation for the truthful study of scientific evolution.

The second term for evolution is "micro–evolution" this is evolution resulting from small specific genetic changes (generally considered to occur over very long periods of time). The small *controlled* changes to the bacteria (as mentioned above) would be an example of microevolution.

The third term for evolution is "macro–evolution" this can be described as evolution on a large scale extending over vast geologic time and eventually producing new and novel genetic material. Macro–evolution was NOT observed in that deceptive experiment described above.

In fact, macroevolution has NEVER been 'observed' by science, it has only been imagined. It's fair and honest for me to say that the practice of using the term evolution to mean either (or both) of these very different types of evolution is absurd.

So what happens when Richard Dawkins is challenged to explain macroevolution? He might say something along these lines… macroevolution is simply many microevolution events occurring over vast periods of geological time. Does that sound like circular reasoning to you? Well it should! Microevolution occurs when an existing creature generates programmed changes to its DNA. So why do I have a problem with that?

The problem is that an assumption is made. The assumption is based on a starting point of one creature (perhaps the first living cell) then microevolution takes place and so on… Such an assumption can only explain the changes to an existing living thing it can never hope to explain its *existence*, in the first place.

The story will go around in circles every single time, and each time it comes back around to count the number of creatures at the start. Each time the evolutionary scientist rolls that idea around in his head, the number at the start still NEEDS to be *one* (or it doesn't fit the mind). However as soon as you start the circle with the number zero, (the actual number present on the sterile Earth), the whole micro–macro fiasco falls down in a screaming heap.

The problem just described is much simpler to grasp for the man–kind, than it is for the scientist. The person still living in the man–kind hasn't been formally inducted into the mysterious world of scientific delusion – I mean evolution. For that reason a person can 'see' that only macroevolution can produce the new and novel information required.

Dawkins' faith in evolution requires a 'finished product' (to work with) and that finished product needs to be in place right from the get go. The very first 'finished product' (by definition) couldn't possibly have had any previous product for microevolution to work on – no matter how much deep–time you throw into the equation! Try to 'see' just how unlikely that 'first event' really was!

Also, while you're at it, try to consider why such a process would have stopped happening as soon as it began. My mind cannot rationalize such a well–timed event. The first living cells would need to know when to stop organizing themselves from non–living matter, so that evolution could take over from there.

What is cause and Effect?

Sunburn is the effect you feel after staying in the sunlight for a long time. The cause is the heat from the Sun.

The law of cause and effect is so universal that it's been granted the status of a scientific law. For something to be established as a *SCIENTIFIC LAW* it must come under rigorous scientific scrutiny. Part of this law's definition states that you may never have an effect that is *greater than* its cause.

Thermodynamics is the scientific study of the dynamics of heat. It is made up of three basic laws, on which (ALL) disciplines of science are based. The Second Law of Thermodynamics is directly related to cause and effect.

For something easy to understand—if you spin a top, it will eventually stop spinning. The energy used to perform a particular task changes from usable to unusable during the process. It will always go from a higher energy level to a lower energy level—where less and less energy is available.

With the second law... everything moves toward a state of disorder—this process of decline is often referred to as entropy. In the case of your 'mobile phone' for example, as your battery declines – its entropy increases.

This is related to cause and effect.

Scientific laws cannot be broken.

Scientific laws cannot contradict each other. Scientifically (using of the Second Law of Thermodynamics), every cause will create a *LESSER* effect!

How would a more advanced life form LIKE SAY A HUMAN, (which is the effect), be generated from a simpler life form, (which is the cause)? The theory of evolution is a field of study where the scientific community can **'subvert the law of cause and effect'.**

The beginning of life on Earth required various forms of evolution (to arrive at the beginning–the first living thing).

I can show some basic areas where evolution could happen, but the full scope of evolution goes far beyond these six requirements.

- **Cosmic evolution:** The origin of SPACE AND MATTER (the observable universe), and the origin of time.

- **Stellar evolution:** Sometimes we think of life as being made of "STAR STUFF."

- **Planetary evolution:** The origin of our solar system.

- **Chemical evolution:** The study of how the various chemicals were formed.

- **Macroevolution A:** Skipped, because before there was organic life there's were ZERO living things to perform 'micro' on, therefore macro cannot possibly occur at this stage (between the chemical and the organic. The only way that (Macroevolution–A) could work (at this stage of evolution) is by using some other method. It cannot possibly use method B because that requires many years of microevolution.

- **Organic evolution:** The study of how the first living organism was formed)? There isn't much support for this one. Remember the requirement for new words to help make evolution more palatable? The word contrived for this stage of evolution is abiogenesis. Basically, this refers to the study of (living–things being generated from nonliving matter). I can assure you that no human has ever witnessed such a thing.

- **Microevolution:** The evolution within different 'kinds' (after they became organic creatures).

- **Macroevolution B:** By now you have noticed that microevolution only deals with mutations *within* a species. Macroevolution B, tells us that such adaptations and mutations allow new and novel DNA to form over very long periods of time–thus producing a brand new finished product (a new species)?

I cannot explain where one particular disciplines starts or where it overlaps. It is quite difficult for me to determine which research or which evidence is related to any particular discipline at any given time. Each one of the above types of evolution got lumped together in an unruly and unscientific way.

There is a great deal of evidence to support microevolution. This kind of evolution seems perfectly valid because it is happening all around us, and it has been recorded by many writers over hundreds of years. Knowing that a virus can become resistant (to antibiotics) is evidence that supports microevolution. But, this evidence must not be used to "prove" that macroevolution has occurred. That would be a *HASTY GENERALISATION.*

When an assumption is used to support or prove other aspects of evolution, it could be called 'begging the question.' The reason why this method is faulty is that the assumption of something occurring is not the same as the EVIDENCE for something occurring.

A scientific theory is a "theory that explains scientific observations" and "scientific theories must be falsifiable." A generalization cannot be tested (or proved to be factual).

When evolution of any kind is described to us, it seems really marvelous, but many of the statements used are deceptive and only lead to another dead end.

A science writer named Martin A. Moe, (*Science Digest)*, wrote, "A century of sensational discoveries in the biological sciences has taught us that life arises only from life..."

The science of living things coming from non–living matter is "wrapped in a mystery." As each theory is debunked a new and improved explanation comes along to take its place, and the cycle continues.

Room for improvement: The story of L–Tryptophan.

The following is an unpleasant story, but it is necessary to show one of the reasons why evolution is so unlikely. Please think this through carefully—it is a story of life and death importance. This story shows why genes should not be modified, not by mankind and not by random changes.

The final years of the 20th century have seen a scientific gold rush – a heady race to make fortunes from genetic–engineering. As with any gold rush there will doubtless be a handful of winners. No one yet knows how long the list of losers will be.

Somewhere on that list will be the name of Marguerite Vitolo, mother, writer, fashion designer… and cripple. Marguerite is one of the 1548 victims of the world's first disease triggered by random genetic modification.

As she lies on her bed, coughing uncontrollably and racked with pain, she has a simple message, for consumers bewildered by the fast–moving claims and counter claims of the GM debate. "NOBODY KNOWS FOR CERTAIN WHAT THE LONG TERM RESULT OF GENETIC MODIFICATION WILL BE," she says. "You allow this food to go on sale, and every trip to the market will be a game of Russian roulette."

Marguerite's frail frame is testimony to what can happen when a worst–case–scenario comes true… She had lived a full and fruitful life between her native New York, where she had an apartment near the Empire State Building, and her Central London flat. At 62 she was slim; fresh faced, poised, and was still turning heads in the street.

Then, suffering insomnia after a car crash, she began taking an all–natural sleeping pill called L–Tryptophan. She smiles slightly as she recalls… "I don't want to take any drugs, and the manager of my local health–food shop kept saying how great these things were." L–Tryptophan is an amino acid that has long been thought to aid sleep and has been used successfully in the treatment of PMT, anxiety and depression. Several pharmaceutical firms HAD BEEN SAFELY MANUFACTURING IT FOR SEVERAL

YEARS, by extracting it from bacteria fermented in huge vats. But in the late 1980s Showa Denko, a Japanese petrochemical giant (which was branching into dietary supplements) decided to accelerate L–Tryptophan production. Scientists at the firm's Tokyo plant produced a genetically modified bacterium (codenamed string V), by inserting several genes that caused it to produce far more enzymes than normal. The company believed the process would cut production costs by about $800,000 per year.

There was a terrible price to pay however. The process also generated a rogue amino acid called "EBT" which is highly toxic. In early 1989 the pills went on sale across the United States, appearing in health–food stores in packets identical to those containing L–Tryptophan made by the conventional process.

Customers shelled out $15.00 per pack of 50, and although the amount of "EBT" made up less than 0.1 percent of each pill, IT WAS ENOUGH TO KILL THEM OR LEAVE THEM PERMANENTLY DISABLED. Soon Marguerite was suffering nausea, flu–like symptoms and muscle pain. Across the country there were sufferers with identical symptoms. Others found their skin was blistering, some suffered swollen livers and lungs, and many complained of memory loss. One of the worst symptoms was nerve damage resulting in muscle weakness, which could be so severe that patients were unable to sit up. Some lost the ability to breathe and were put on respirators. Others developed huge non–cancerous growths inside the chest or behind the eyes. Marguerite says: "I felt as though some kind of alien had invaded my body. I had a whole battery of doctors around me, doing countless tests, and they knew I was ill because they could see my immune system was going haywire, but none of them had a clue what was wrong with me." By the end of the year she could no longer write, she was forced to give up her apartment because she could not walk up the stairs, and her weight was falling rapidly.

Faith Rumph, 49, a piano teacher from the small town of Dumfries, Virginia, started taking L–Tryptophan to combat sleeplessness on the recommendation of her doctor. Today she cringes as she recalls the livid rash that spread across her body. "The skin began to thicken and go a yellowish brown; it was scaly like a reptile. My legs were the worst. The hair couldn't grow through because my skin was so tough." The skin on her fingers became so tight she could no longer play her beloved piano. Her joints began to ache, and then her skin simply started to peel away in large slices.

"I was in absolute agony," she says, "I couldn't stand wearing clothes. My scalp began to fall off."

Both women were eventually diagnosed as suffering from eosinophillia myalgia syndrome, or EMS, caused by a sudden rise of the number of eosinophils, a type of white blood cell that usually protects the body from parasites. The victims of L–Tryptophan were mostly women in the 40s and 50s age group. Because they were spread thinly across the United States and Canada it was several months before their doctors began to realize they had all been taking the same 'natural' sleeping pills. The breakthrough in understanding the link between Showa Denko's "all natural' dietary supplement and EMS was achieved by doctors at the MAYO CLINIC in Minnesota. The US Food and Drug Administration ordered the immediate recall of the company's L–Tryptophan in 1990. [Coffs Regional Organic Producers Association Inc., "CROPO" – March 2000]

Yes I understand that the story is meant for those who would consider Genetically Modified Organisms. But I want you to consider the details from an evolutionary point of view. This story completely defies the idea of nature making random changes to the genome. The changes made to the product shown above were planned and designed (not just random changes). The truth is, if dear old mother—nature had made just one mistake, over millions of years, it would have eliminated the human race for good.

But do we see nature making changes like that? What if the basic pattern for life (DNA) didn't change much at all over a period of 540 million years, would that be surprising to you?

Chlorella: sometimes known as "Sun Chlorella."

Chlorella is a microscopic, (freshwater), single celled, GREEN ALGAE. Modern science has discovered that Chlorella was alive and well 540 million years ago, so how do they know it was chlorella? Because its DNA pattern (the pattern of the ancient chlorella) is almost identical to the DNA pattern of chlorella grown as food in our time.

Remember DNA can be used to identify your close relatives. Chlorella would be a perfect example to become one of the candidates for the early theory of pond scum. The only problem (for the tree of scientific evolution) is that the pond scum didn't change significantly over 540 million years. Now think it through, the reason why humans can eat chlorella as a health food supplement is because its DNA is compatible with the processes of the human body. If the chlorella had taken even one (wrong) turn over 540 million years it would no longer be a viable food source for any living thing. Put simply, the one (bad) change to the DNA of chlorella would have had an effect on all living things, all the way through the food chain. In other words, life as we know would not exist in this present age.

The single celled "super food" contains 60% protein 18 amino acids and a number of vitamins and minerals to boot. One of the properties of chlorella is a plant—derived nutrient called Chlorella Growth Factor, C.G.F., which is a nucleotide peptide complex found in the nucleus of the chlorella cell.

Evolution science makes a great fuss over the shared DNA of various creatures and then they use that information as the scientific basis for arranging and teaching the 'evolutionary' tree of life.

What they don't teach is that if the DNA were being randomly modified over long periods of time, the planet would have ran out of food millions of years ago.

The fact that any creature is still alive (to this day) is living prove *against* the theory of evolution. If you are living in the mankind, you can think that through and make up your own mind. On the other hand, if you are an atheist evolutionist scientist you must overlook the information that has been provided, or find some way to side–step the profound implications.

At the center of the idea of improving a species or organism, is the assumption that mutations will produce more and better traits or characteristics. To produce more and better, requires *new* information. The place for all that information is of course the DNA. For new information to have formed there would need to be an injection of the said information by some sort of mutation. But as you can see (from the sleeping pills) the new information is troubled by serious and very real problems. The same sort of problems must occur whether the change is made with a purpose or whether the change is randomly generated.

Genetic mutations are spontaneous (chance) changes, which are rarely beneficial, and more often have either no effect, or a bad effect. As the information is observed from a truthful perspective evolution becomes even less likely.

Changes of the 'evolution' kind are extremely unlikely to produce a single new and improved species, even if the changes happened over millions of years.

While science is allowed to fabricate stories, they do not like it when humans take that approach. Lucky for me I am not under the rule of science I am free to tell a story – whether *they* like it or not...

Try to imagine a small group of hardened evolutionists gathered outside the establishment building. They are there to dictate terms to another small group from the man–kind. Then along comes Charles Darwin in person.

As soon as a scientist–spokesman arrives at an impossible part of the theory they would automatically fall–back to the chant of "Darwin, Darwin…" But this time, a person of the human kind rushes over to grab Charles by the arm, and tries to drag him away from the scientist.

The loyal followers of all things Darwin soon become angry. One of them grabs Darwin and drags him inside the building. Shouting: "this guy's with us."

Within minutes Charles Darwin speaks out. "Those humans out there – they were my friends. They were willing to speak out about my work. I stand with them! Those are the ones who accepted **that the theory of evolution has been annihilated.**"

You should be able to guess what happens next…

The scientists by now have become an angry mob. Still a bit annoyed because they have failed to hypnotise the crowd of humans outside. Now their saviour has turned against them as well. The only thing left for this group of scientists is to bundle–up Darwin and eject him from the building. Then with Darwin out of the way they can go back inside and continue their work on 'Darwinian evolution.'

In future, every time you hear a representative SELLING Darwinism I want you to recall what science has done against his theory.

Whatever you do, don't try to argue the point with the person who preaches Darwinism. Always remember they are just following the terms and conditions of the scientific establishment. They cannot change their ways, even if they wanted to, because such a change toward truth would land them outside the boundary of scientific naturalism, (in more ways than one).

Chapter Seven

MORE INFORMATION PLEASE

The process of evolution cannot continue to improve a species, by changing or losing information. The reasoning is quite simple, over a long period of time the constant loss of genetic information would lead to no information at all.

The only way a new species could be formed is by the production of new and novel information. Information would need to be added to the DNA strand in a very precise and orderly way. There can be no exception to this! Random changes find it very difficult to be precise and orderly.

Remember the early induction into the mysterious world of evolution. In any population of organisms that can be observed there is a range of variation. This variation always has a limit, (it can only go so far). Keep in mind that mutations are mostly either harmful, or neutral.

When a change occurs there is an internal mechanism that will repair or reject unwanted changes, but how would the mechanism sort out helpful/destructive changes? The answer is – **if it were truly a random process it could NOT distinguish between the two.**

One of the evolution 'things' I learned in school was the story of Darwin's finches. We were taught that birds with different beak sizes eat different foods. Because of this, they live and breed in a different location. Or perhaps that would make more sense the other way around.

In the process of learning, we were shown some photos of birds that were all finches, yet they had distinctly–different beaks. These differences were taught as evolution. What we were NOT taught is that the changes were occurring in a cycle, and that eventually (over generations) the beaks of these finches would revert back (to a starting point). The FACT that it was a 'cycle of change' should have dismissed any thought of evolution in action. But it didn't!

Another fact that arises from the study of finches is that the information for beak size is poly–functional. Birds with short beaks also have small feet. Birds with long beaks have larger feet. This proves that the information stored in the DNA strand is used for more than one purpose.

This causes a serious problem for the idea of random improvements over vast periods of time. A random change that improves one aspect could easily ruin something unrelated. Unintended changes could lead to chaos rather than long–term improvement.

A few weeks later we were taught the story of the peppered moth. The Peppered Moth was another 'thing' that should never have made it into the science textbooks. It is a little lengthy but please read through it. The idea of this being an example of evolution–in–action was absurd in the extreme, (even while I was still at school).

The story, from evolution–science, starts back in the 1850s and the study occurred in the north of England. It was a time when there were light and dark moths in the moth population.

Before the industrial revolution, lichen on tree trunks was a light colour. The moths would (supposedly) rest on the tree trunks during the day. Birds (predators) would come along and feed on the darker moths because they were easily seen. This process over time, led to more of the lighter variety in the population.

One hundred years later, the work of industry had blackened the tree trunks with soot. By then the lichen was long gone! This change over a very long time, led to more of the darker variety in the population. The decline in the lighter and increase in the darker moths was being taught as evidence for evolution in action.

It was right there in the science textbooks!

But what happens when you think it through?

After 100 years of change the area still contained a population of light and dark peppered moths. So what is the point of all this?

Well, even if the lighter moths in the population were finally weeded–out altogether, you would never have thought of that as an example of evolution in action. If one variety dies out completely, that could only mean that the DNA for that *variety* has been COMPLETELY WIPED OUT. Evolution cannot succeed into the future by wiping out information!

In the above example the peppered moth starts out with variety and diversity and ends up with less variety and less diversity. Yet for generations science students were shown this material as an example of evolution in action, in reality it could only be described as evolution in reverse! To make matters worse it is highly unlikely that the moths even rested on the tree trunks during the day.

Imagine if I was inside the scientific establishment but remained a member of the man–kind. If I wrote about the finches implying programmed switches, with a pattern for the bird to develop a different beak, I would not get any peer support to continue with that line of work.

If my work slipped through the cracks, I would point out that there were multiple patterns 'built in.'

I'd write about a type of change that is pre–planned, and does NOT support the scientific theory of evolution. From the idea of a pre–existing pattern, I could then look for another example of a 'trigger' occurring in nature, another pre–programmed event.

The Axolotl already has the 'pattern' to develop into an adult, but it doesn't progress. Yet, if the Axolotl could be artificially moved into waters that contain the right essential–elements, the 'baby' will grow into an adult salamander. The change into an adult isn't a form of evolution, because the pattern for printing the salamander MUST be available before the event is 'triggered.' So the Axolotl is patiently waiting for an event to occur, it's waiting for the water to 'gain' the exact material that is required for the transformation (to be successful). If the Axolotl didn't have a mechanism to (know) what it is waiting for, it would never make that change and *we* would never have observed the process.

The main point here is that the axolotl doesn't just randomly decide to change. The timing of the event has been preset. The second point is that the trigger would have caused a catastrophic event unless all the code was in place – ready for the new and novel creature to appear. This cannot ever be thought of as a random process, it is obviously a pre–programmed response with a pre–programmed set of instructions for building the finished product.

Students trusted that scientists and teachers understood the work of Charles Darwin. They trusted that their peers were capable of discerning fact from fiction. But unfortunately evolution has become a popularity contest held together by shaming their opponents.

Working in science should NEVER be reduced to a popularity contest.

For science to earn back some credibility (in my world) they need to urgently return to teaching only the factual information. If they must continue to teach evolution as if it were fact, they should at least stop the practice of including the word "Darwin." He was smarter than that!

The missing–links of scientific–evolution come and go like the tide and the explanations of how particular 'episodes' of evolution came–about, are continually modified to suit the need.

The stories of how amphibians conquered the land, how the birds developed wings or how the dinosaurs became extinct are all just stories, nothing more. They are just grubby speculation repackaged as scientific thought. The idea that humans evolved from apes is just a product of the 'science,' motivated by the need for a naturalistic solution. The study of genetics shows that apes and humans when compared have **a large number of biological differences**.

Students tend to trust their teacher, (at school and in science later on). Yet the method used to teach evolution is effectively the same as brainwashing. Propaganda is the main supply of information. Teachers must present evidence that supports the science, without the need to demonstrate how the evidence fits the outrageous claims.

Think about your time at school. Were you ever taught that **actual experimental evidence,** demonstrating a known case of macro–evolution is **NON–EXISTENT?**

Chapter Eight

FROM MS-DOS TO WINDOWS

Yes I know, it's hard to believe that Richard Dawkins presented this 'evidence' in "The Virus of Faith" but he did. That one small blunder (on his part) is enough to make us all suspicious of **his** 'scientific' theories.

In my introductory book, I put forward an idea for the science teacher. I asked the question (What did the computer have going for it that the monkey's cousin couldn't possibly have had). The answer of course is MAN–KIND. Remember, the word mankind is all about the kind of creature you are.

An existing living being (a member of the man–kind) designed and produced the computer.

In this chapter we take a brief look at computers and their requirements and the various stages of development.

With some thought the development of the computer can be compared to the development of the man.

The dodgy claims about the fossil–Lucy have been covered elsewhere in this book. By the time you are finished this chapter you should come to realize that making such an outrageous comparison was a very foolish (non–specific) thing to do!

Computers didn't evolve on their own, yet the comparison "wonders" if they did? The other glaring mistake (when it comes to the theory of evolution) is that the ms–dos–computer was a finished product and so was the monkey's cousin.

Evolution needs to get away from concentrating on finished products and begin explaining the slow and gradual process that must have occurred. The story of Lucy's foot (and the mathematics of it) is vitally important and you should bring that to mind whenever you feel the need to question this. (Mathematics doesn't lie), scientifically speaking there MUST be either evidence for gradual change or evidence for random change (or a combination of the two). The only evidence that Lucy provides is evidence for a one–off change. A foot morphing into a new and novel foot (in one go) is evidence *against* evolution. It is the exact opposite of what the scientist is attempting to prove, which you should find very strange indeed.

You might think that I am waffling on far too much about the problems for evolution. I do this because some readers will have believed evolution for most of their lives. Such a strong and long–held view is almost impossible to break!

For the remainder of this chapter I am going to compare the progress of the computer with the existence of a living–being, as (thought out) by Richard Dawkins.

As you read this material try to keep in mind that the computer did not make itself, and it did not improve itself.

The ms–dos computer was mostly a command–line computer, and because of that it had no idea that it was an unfinished product. Just as the 'pre-Lucy' of chapter five had no idea that *it* was an unfinished product.

The pre–Lucy was in fact a finished product that already had the capability to survive. It was surviving the conditions that prevailed at that time – there simply wasn't a single reason for a change in pattern and no hint of a mechanism allowing for that change to occur.

The most obvious thing about the comparison (arranged by Dawkins) is that it starts with an already completed species. However, to show respect for the intelligence of the man–kind we need to start the story of the computer from the beginning (rather than from a finished product).

At the time of writing all commonly used computers rely on a counting system called binary (or modulo 2). This system must be used because the machinery of the computer can only count to one. (The machine can only 'understand' two different states).

- A state where a value is returned (usually 1)
- A state where no value is found (usually zero)

WITHOUT a human to BUILD the computer (to calculate using ones and zeros), the computer could do NOTHING

WITHOUT a human to CODE the computer (using ones and zeros), the computer could do NOTHING.

The whole system needed to be designed to work–with the available knowledge (the ones and the zeros).

So how does that system compare to the suggestion of the–monkey's–cousin from millions of years ago?

The monkey's cousin already had pre–programmed hardware and software – operating smoothly together as a unit. Not in binary code (as invented by man) but in modulo–four code. The DNA strand inside a living thing would come to be known as the most densely packed collection of information ever observed.

So how do we know for sure that it is coded in modulo–four, and what difference does that make?

Well the most important difference is that modulo–four is exponentially more efficient than binary. We know that modulo–four is used because of the way the parts–list appears within the code.

- Complex proteins are manufactured (inside the cell) using a specific order of carefully chosen parts.
- There are 22 different parts (each with its own unique shape). [There would be more than 22 but the ones shown are the ones likely to be involved in manufacturing the body proteins].
- The correct sequence of part numbers is stored in the DNA using combinations of only 3 DIGITS. (That's the proof for Modulo–four). In man–made computers, if you were using only three digits, the machine could only count to 7. To store 22 unique numbers (in binary) would require at least 5 DIGITS (see the chart below).

As you can see, this is only a small value, yet the gap is already widening. The *old* code, (from millions of years before the digital revolution), is already exponentially more efficient than anything that man has ever designed.

Here is an example of values, comparing modulo–four–counting to binary–counting.

Part Number	Modulo–Four	Binary Coded
0	000	00000
1	001	00001
2	002	00010
3	003	00011
4	010	00100
5	011	00101
6	012	00110
7	013	00111
8	020	01000
9	021	01001
10	022	01010
11	023	01011
12	030	01100
13	031	01101
14	032	01110
15	033	01111
16	100	10000
17	101	10001
18	102	10010
19	103	10011
20	110	10100
21	111	10101

To see how the math works in this table – note that
the value 20 can be seen as 16 + 4 on either scale.

In practice letters are used for modulo–four values. I have used numerals in the table above, so that the base four counting is easier to 'see.' The imaginary "first living cell" needed these base four instructions (already in place) in order to build itself. The first living cell *must* have at least some of these parts ready to rock and roll (or it fails). The first living cell required the parts to be assembled in a precise order, right from the start, that's what stops abiogenisis from becoming a scientific theory. That's the gap in the circular reasoning associated with evolution. The information I have given above shows very clearly that, lots of little bits of micro–evolution could NOT have created the first living cell.

The term macro–evolution is described by science as lots and lots of micro–evolution, occurring over long periods of time. That, by definition requires some living material to do–the–micro on. Before the first living cell there were ZERO living cells available to work with. Without a living cell for micro to work on there can be no macro–evolution. Yet (to some of us) it is quite obvious that a macro–evolution event was definitely required to produce the first living cell. MY QUESTION TO YOU IS SIMPLE... WHAT PART OF THE ABOVE DOES SCIENCE NOT UNDERSTAND?

To see what I mean about circular reasoning, think of a clock face, but instead of the number 12 at the top, place the value zero at the top. The science of evolution starts with the number 1 then micro–evolution takes it step by step all the way around to the 11th living thing. (This is the way their science teaches the mathematics of evolution). You and I are entitled to think. You and I are entitled to question the scientific theory of evolution. You and I are entitled to ask... WHAT HAPPENS WHEN YOU GET BACK AROUND TO THE ZERO?

For me, the zero is very real. For me, the circular reasoning just keeps going back around to the zero and there it comes to a grinding halt. Circular reasoning must never be used to 'sell' a scientific idea. Science must be done without the nonsense. Think of the ridiculous experiment used to induct humans into a belief in evolution. In the lab a micro–evolution event is shown and then the known outcome is used to support a belief. That belief system gives the inductee a faith that lots of micro (over time) will make macro–evolution occur.

If there had been a genuine scientist present in the lab, when that false demonstration was used as a tool, he would have explained the true situation. There was zero evidence for 'evolution in action.' The only evidence shown was given to prove a pre–programmed response, which is almost the opposite of 'evolution in action.' Because it is a pre–programmed response it is predictable and therefore it is a repeatable experiment. If it were a random event as taught by science then it would not be predictable, only predictable outcomes could be used as a false tool in this way.

I hope you can see, that when science (or anyone hired to promote science) tries to 'sell' the theory, by showing things like the monkey's cousin, they are being willfully deceptive. The monkey's cousin represents a finished–product. There is zero evidence to support the existence of the monkey's cousin (by a process of scientific evolution). Using such a thing to support a failing theory is hardly a scientific approach, whether you call it evolution or Darwinism or Neo–Darwinism or some other device makes no difference, impossible still means impossible.

Now with that out of the way we are ready to take a good look at the origin of *man–made* computers.

The Math:

The first 'event' leading up to the existence of computers was the invention of a digital counting system.

The modern binary number–system was proposed by Gottfried Leibniz around 1679 and shown in Explication de l'Arithmétique Binaire (published in 1703).

The first living cell (by definition) required the invention of modulo four, but (millions of years ago) there was no–one around to design it. Modulo–four (according to the theory) just happened by chance, as part of a random process of gathering information (and bringing it to life).

The Energy:

The second important 'event' leading up to the invention of a binary computer was the discovery of electricity. For a good understanding of the differences in the energy supply we need to consider the energy at the molecular level. This of course would require a whole book (to do it justice), however it is sufficient to say that the energy humans harvest and transmit (to run their computers) is very archaic.

The energy of our world, the energy that we convert and transmit and use (to power the computer) is electron energy. The strong energy of a molecule is the PROTON and the weak energy is the electron. Note that the *charge* of these particles is similar, but the *mass* is different. The energy system that was planned, designed and built by humans uses the weakest part of the molecule. Yet (by chance) the energy system inside our bodies is exponentially more efficient. Our internal system uses the strongest part of the molecule. We run on proton energy!

Inside our body, billions of tiny machines convert proton energy into a form of 'energy–currency,' allowing the energy to be used where it is needed throughout the body. The scientist must wake up each morning and convince himself, that these microscopic–proton–energy–devices–inside–our–bodies were produced by some unknown event in distant the past.

The Device:

Once we have a system of math that works and some electrical energy, then we can build the device. Yes, we take these devices for granted these days. The only time we realize they are man–made is when they fail. We know from the math that the human body is superior. We know from the energy supply that the human body is superior.

So what about the scientist of the future? Fully convinced that he has seen evolution in action, he decides to take a field trip. During his travels he comes across a site where human remains are being uncovered. At the dig he sees a human–hand clutching a mobile–digital–device. This is the point where his mind begins to squirm (like a bowl of noodles). At this point he MUST believe opposing truths.

- He knows that the technology inside the human is vastly superior to the device.
- He knows that the device had a designer.
- He knows that the device had a manufacturer.
- He must believe that *our* internal technology had no designer.
- He must believe that *our* internal technology had no manufacturer.
- He must continue to believe that the technology inside the human body came about by chance, over vast periods of time.

Science was never meant to be a faith based system, yet faith in evolution requires a belief that the human body somehow got to the finished product, and somehow became better than anything *we* could ever hope to design or build.

What you see:

One of the things that made computers popular was the video screen. The early monitors were called monochrome (usually a green or orange tone). Then came colour screens. The first colour display had 16–colours on a small screen. Next 256 colours, and a slightly larger screen. Our human vision was ready for these changes. Our human vision made these images viewable in the mind. The modern display has thousands of colours available and thousands of pixels to display them on.

Videos, when displayed on a computer screen have a frame rate of approx. 24 frames per second. For 3D viewing, a duplicate of each frame is used. One frame is passed through a red filter and the next is passed though a blue filter. Yes there are other versions of 3D, but this is a common one at the time of writing. Showing 24 frames in 3D requires 48 frames per second. Each and every one of these frames is packed with information. Our energy system, our eyes and our brain must work in concert (to allow us to watch the concert).

Multiple frames per second, each one packed with information and our eyes automatically turn it all into something we can see and understand.

> As the computer screen advanced (over the years) our eyes have kept up. In fact our eyes are still ahead of the curve. Our eyes can do better than 48 frames per second.

> Yet, according to the theory of evolution our eyes are still pretty much the same as they were millions of years ago.

Does that make sense to you? Did the monkey's cousin need HD vision? If you are a scientist you are not required to think about this. But what about me, I am a member of the man–kind. Do I have the right to think about where my HD vision comes from? I believe that I do have that right. I have defective vision, yet I can still see the difference between a 640 x 480 screen with 256–colours and a HD screen with thousands of colours.

We all know that closing your eyes for a few minutes is very relaxing. Have you ever wondered why?

Your eyes use more than half of your brain's available processing ability. Closing your eyes makes the processor available for other tasks and gives your energy system a chance to recover. When driving a car you must concentrate on the road ahead, you must keep your eyes open. That is why driving is so tiring, that is why you need to stop and have a rest. Up–close video monitors cause more stress on our eyes than driving a car. When you watch something on your phone or tablet you are focusing on an image up–close, this requires extra energy, it also causes eye–strain from the constant focusing. The eye–strain will eventually lead to eye problems later in life.

Whenever you are watching a long movie, show it on a TV or on a large monitor, and get as far away as you comfortably can. Watching anything up–close will strain your vision. If the movie is boring (and it usually is), switch the TV off and go outside. Your eyes will work better outside where there is natural light and you are looking out over a long distance.

The sound:

The computer, over time has gone from no sound, to mono sound, then to stereo, then to multiple channels. Our hearing was already ahead of these inventions. When a sound is behind and to my left, I automatically hear it from that direction. Our hearing ability hasn't changed much over millions of years, yet it doesn't need an upgrade for digital sound.

The Printer:

Being able to print stuff out with your computer makes the computer experience more enjoyable. We now live in an age of 3D printing. The scientist thinks of the 3D printer as a massive advance in modern technology. What they don't want to think about are our human 3D printers.

Inside your body there are thousands of 3D printers manufacturing human hair. These printers work day and night "growing hair" for us. The rate of 'growth' is around 1^{mm} every two days. If you don't eat well or if your body is under stress then you cannot provide the right 'ink' and the printers stop producing hair. One of the ingredients for the ink is a B vitamin known as lecithin. Most human diets are deficient in this vital ingredient. When your supply chain runs out (of lecithin) all new hair will be printed without natural colour.

The flour that humans eat as food is a left-over product (anything worth eating has already been removed). You must remember that eating foodless–food is the first step towards a decline in hair quality.

The first thing you notice about hair production is that hair and skin and blood vessels form a very complex structure. This structure is called the hair follicle. Hair has a natural cycle, when one dies–off another begins growing in its place. The hair 'grows' and dies off in a long cycle.

- The growing phase begins by pushing out the old hair and then continues for up to six years.
- When growth slows, hair production enters the dying phase. When the printing stops the follicle is broken down into basic parts and recycled for use in the body. This part of the process may take weeks to complete.
- The resting–phase that follows lasts for about six months. During this final stage the hair is not attached to the machinery and it loses condition, and can appear dull.
- Signals from the scalp and other parts of the body control every step of the hair printing process (year in and year out). Some signals tell the hair to die and fall out. Other signals tell the process to start building another printer and start "growing" hair all over again. These are called growth factors. Then the same signal is used for the length of the growing cycle.
- When the signal changes again the printer is broken down and recycled, and so on.

The hair growth cycle is a complex process. If the process had been put together at random (using trial and error over a long time) it could not possibly be so efficient. Take away just one small part of this cycle and everything changes.

Changing just one thing would see all hair follicles working together on the same signals. So what?

You would have nice hair for a while, and then it would look fairly dodgy for a period of around six months. After six or seven years it would all fall out at once, (as soon the new printers were up and running).

When you are first born you have more than 90 thousand scalp hair follicles. This number varies according to your hair type and colour.

- For blondes, around 150 thousand.
- For redheads, around 90 thousand.
- For darker hair, around 110 thousand.

These numbers are just for the scalp area. If you could count the entire body it would be many more than that. Each and every one of these hair follicles is a fully functional 3D printer, working in a programmed environment. This causes a weird situation for the scientist.

The evolution scientist is forced to believe that this hair process was in place and working reliably (on a 6 to 7 year cycle) millions of years before humans first even thought about building a 3D printer!

I don't understand how they keep it all lined up. Humans have discovered 3D printers working on a controlled time cycle, and powered by proton energy. A printer that is manufactured inside the human body and kept going by a computer–process that operates using modulo four signals. This signaling and printing process requires technology. The technology goes way beyond *our* feeble attempt at a 3D printer, yet it somehow made itself by chance over vast periods of time!

Do I really need to go any further?

It was a bad idea to say that some human trait can be compared to windows. It was a bad idea to say that some monkey's cousin was the starting point. Richard Dawkins should have sat down and contemplated what he was doing, before he got someone to bolster his faith in evolution.

The monkey's cousin was a finished product. The human body is a finished product. There is no visible evidence to support the upgrade occurring over vast periods of time. Think back to the math of the human foot and howl with laughter. I used the foot as an example because it would be impossible for evolution to "monkey around" with the foot without leaving a trail of evidence. There are millions of fossils in the fossil record, yet there are no convincing examples of any poorly made or (failed) feet, only a human like foot, and that foot (according to scientific evolution) was already in existence millions of years ago.

The foot hasn't changed its pattern much for millions of years. The apparent cousins of the human haven't changed *their* pattern over millions of years. So what mechanism stopped this process from re—occurring, after a result was obtained? Why did evolution stop suddenly? How did it know when to stop? Did the pre—monkey's—cousin and the pre—human both know that the correct foot had been produced in one go, (in order to suddenly stop making changes)? Evolution scientists need to explain how that fits with the slow – and – gradual – change – over – time model that they are so comfortable with!

Chapter Nine

THE WRONG DIRECTION

The laws of science imply that anything left to its own devices will decline into a state of disrepair. The laws of scientific evolution rely on a belief that living matter is able to defy the laws of science. Living things are given a free ride (by the establishment), for the sake of upholding a strong belief in evolution.

I had the misfortune of falling into a lengthy discussion with a hardened atheist scientist regarding the broken theory of evolution. This person was a relative of mine, for years he dictated terms to me and *pretended* to note the plain straightforward facts that I was presenting. There was a logical and almost friendly discussion going on, until I got a little too close to the truth. One day I must have broken down his wall of science, I must have got–to–him, (as they say). That was when he chose to play the 'science' card.

He said, "I am a scientist and you are not, I don't have to discuss this matter with you at all. I was only communicating with you because you are a relative."

Does that sound like a scientific approach to you? His confidence in Darwinism was soundly based on many things.

- His faith in Darwin (not his scientific work).
- His belief that he had witnessed evolution.
- His belief that science has great support for the doctrine of scientific evolution.

When I informed him, that the view of Charles Darwin directly opposes the view of The Scientific Establishment, he confidently assured me that I was wrong.

The scientific establishment is not interested in the work of Charles Robert Darwin they are only looking for a scapegoat for their bizarre belief system.

There are two parts of science where the laws of logic (and common sense) are ignored. One is the carbon science, where anything to do with climate must begin with the word Carbon. The other is the generating of life from non–life, where everything must end with "isn't evolution wonderful."

You already know how the majority support is policed and you know they don't even have to explain evolution to the lesser species, (those of the man–kind).

This person was pretending to be a fellow human, just long enough to convert my thinking over to his way of thinking. That is exactly the way that a cult leader would work on his subjects. That is not the way that a truthful and honest person–of–science would operate. The moment he realized that I couldn't be swayed, he showed his true method of operation, he became a separatist. This person had not studied the evidence for and against the theory. He had not read any of Darwin's work, yet he was filled with confidence that he was a superior being.

It is bad enough that any scientist would speak in such a way to a human being, but far worse when they turn against their own kind and dismiss *them* for not 'behaving themselves...'

Chief education scientist dismissed:

For denying evolution and global warming

Written by Shaun Doyle...

[Shaun has a degree in Environmental Science with first class honours, specializing in Soil Science—as well as a Graduate Diploma in Natural Resource Studies (both from the University of Queensland in Australia)...]

Dr Gavriel Avital, former chief scientist for the Israeli education ministry, is the latest of a long series of high-profile scientists 'Expelled' for voicing skepticism at evolution. [Published: 19 October 2010]

It seems so harsh! But that is precisely how skepticism of evolution and anthropogenic global warming are treated in modern academic, political and social discourse. No one can expect a fair hearing in the public square any more. Any who hint at opposition should NOT be given a hearing.

Dr Gavriel Avital is the latest of a long series of high-profile scientists who have voiced skepticism at these sacred–cows of modern academia and has paid the ultimate professional price—**termination.**

The case,

Education Minister Gideon Sa'ar has dismissed the chief scientist of his ministry, (Dr Gavriel Avital) over his statements denying the "fundamental tenets of science"—evolution and anthropogenic global warming. The news reports are quite straightforward; there is no other cited reason for his dismissal.

Avital has been an outspoken critic of evolution and anthropogenic global warming during his time as chief scientist, and has attracted the disapproval of numerous scientists in Israel for it. In February 2010 two Israeli Nobel Prize laureates issued an open letter to Gideon Sa'ar saying;

"We don't see any alternative other than to replace Dr. Gavriel Avital with an individual suited to fill the position, one who could do so faithfully and professionally."

Lest any think it was because of poor performance that these scientists wanted him sacked, the letter went on to state:

"We view Dr. Avital's remarks gravely because they undermine the standing and importance of science and take us centuries backward, even as the world celebrates the importance of Charles Darwin's discoveries and the great contributions he made to human knowledge and scientific development.

I find remarks such as this perplexing. I wonder what the well instructed Newton – Kepler - Pastuer – Lister – Maxwell – Joule – Kelvin – Steno or Faraday, would think of such comments. All but three were contemporaries of Darwin, and a number were very public in their criticisms of his work.

His personal doubts about evolution are well known, but what exactly did Avital propose as the *educational* solution? He said:

"If textbooks state explicitly, that human beings' origins are to be found with monkeys, I would want students to pursue and grapple with other opinions."

He's **not** talking about **not** allowing evolution to be taught—he *assumes* children will be taught evolution because it's in the textbooks. Rather he simply wants to give children the right to hear other views. Avital is not the one suppressing alternate views—evolutionists are.

Evolutionists' tactics to silence their opponents

Evolutionists are in the habit of bullying and forcibly silencing their critics.

'Confessed–evolutionists' Michael Reiss and Richard Sternberg were dismissed from their posts because they hinted that people should be able to think for themselves about evolution.

Whenever anyone tries to discuss the evidence, typical evolutionist responses are: "all true scientists believe in evolution." Such 'arguments' are just more smokescreens evolutionists use to avoid dealing with the actual arguments.

"All true scientists believe in evolution" presumes a self-serving definition of "Scientist". It also presumes that majority vote decides truth. As evolutionists are so fond of pointing out, (after Bruno's time) Copernicus and Galileo were almost lone soldiers fighting the tide of geocentrism. And yet we know today that the majority was wrong and Galileo and Copernicus were right. But this is a double-dip — evolution cannot be proclaimed right, just because it has the majority vote. Scientists haven't learned from the history of popularism. The popularity of geocentric thought had zero control over the data. Our earth is not at the centre of the solar system, it is following the sun like a puppy on a lead as we zoom through empty space at almost 600 kilometers per second. In this age, it is logical to assume that evolutionary scientism has no control over the data.

Belief in Darwinism (against Darwin's own words) is a form of popularism that must eventually follow past failures.

For a similar reason, the appeal to "peer-reviewed literature" holds no water either. Evolutionists *are* the "peers" who review and edit such journals.

Evolution is not a scientific question, it's historical. Experimental science needs to make the historical reconstructions believable, but the reconstruction itself is only ever based on some form of naturalism.

Evolutionists play games with the term "science" (whether knowingly or not) because the public views "science" as reliable. Science is a very useful way of investigating the world, but *evolution* is not supported by data (unless the data is supported by assumption).

These charges have been answered time and again, and yet evolutionists generally respond merely by putting up the same smokescreens [they are *not* legitimate arguments], or by actively seeking to discredit and silence dissenters. There are no attempts at rebuttals, because they'd have to acknowledge that there are arguments to rebut.

Why such language and tactics? Why such arrogance (such self importance)? They assert that the truth of evolution is so obvious that it's *impossible* to honestly disagree with it, (if someone understands it). But in the mankind we say, the more you understand the worse it gets!

The emotive sentiments of the scientific mind are communicated to us by many means. Not just because evolution is assumed to be true. It's much more than that. Richard Dawkins says it most bluntly:

"It is absolutely safe to say that if you meet somebody who claims **NOT** to believe in evolution, that person is ignorant, stupid or insane!"

Such a worldview ignores the history of science.

I think it is very wrong for the salesmen–of–science to promote scientists (in general) as people of a higher species.

I learn something new every day. I GAIN KNOWLEDGE.

Just because I don't have a few letters after my name, cannot prevent me from doing science. To pre–fabricate the idea that a particular version of science is wrong, **without consideration**, and then pre–crackpotising the author for good measure, **is hardly a scientific way to deal with the data!**

Chapter Ten

THE AGE OF THE UNIVERSE

I feel that the arrival of a planet, a place for life to exist, helps to define evolution, by finding a starting point for the essential conditions that would be required for life.

This chapter is not directly a matter of organic evolution. It is **one of thousands of differing views** about our physical universe. It explains that our galaxy (and of course the entire universe) may be much older than you think. If you decide to skip this inset chapter, please keep in mind that before the elusive magic of abiogenesis can occur, there has to be a life supporting planet for it to occur on!

There are many things that are known to have occurred, to build a life supporting planet. A good planet needs a galaxy to hold it in place, even better, a planet with a moon, (to provide a tidal system). The heat energy from the sun, and the cycle of the moon are needed, because they drive the 'natural' cycles of the planet.

According to science this good planet was in existence for many billions of years before it became suitable for life. The reason why this section is so important is that we are defining a starting point. The honest and truthful starting place for evolutionary science is a sterile planet, even if that is an uncomfortable position to take.

Most of this material has been covered in scientific articles over many years. The only difference this time is the process of LOGICAL ANALYSIS. Would you believe that logical analysis suggests that the universe could be **much older than it appears to be.**

A great deal has been said about the size of the universe, and when logic is applied, the actual **age** is not found by observing the **size**. To work out our place using the mathematics of the fourth dimension sounds quite difficult at first, but it isn't difficult at all, it's just logical. Calculations for the fourth dimension can be done using two–dimensional thought, and two–dimensional math.

To get an idea what is meant by space–time, consider another sun (close) to our system—say 600 light years away. If you could travel instantly to that visible point (in distant space) you would not be anywhere near that distant sun, because that sun would have been traveling *away from there* for the past 600 years. So what does that mean in space terms? Well, after you arrived at that point (where the light originated)—(even if you could arrive there in an instant), you would still be a very long way, away from the physical object itself. From where it WAS, if you could tear–off in a rocket and immediately start traveling **at twice the speed** of that object, it would still take another 600 years of space travel just to catch up with it!

Now we get to the interesting part. What about the age of the universe? What if the universe was much older than we expected? That's a pretty big call, so let's use some logic to take us further back in time. Please bear with me through the thought process, the calculations will seem complex at first, but once you get to a certain point, it will become remarkably easy to understand.

o   The early calculations of science show a universe that is more than 13 billion years old.

○ There has been a recent discovery of a (young) galaxy approx. 13.8 billion light years away.

○ A rough estimate would make that young galaxy 'appear' to be 700 million years old.

○ The current velocity of our galaxy is approx 580 km per second through open space.

The reason why appearances can be deceptive is because light–speed is different to galaxy speed! As we rip through open space at 580 km/sec, the light passes us at 299 792 km/sec. The difference in these two speeds can cause some confusion about how FAR we have traveled and how much TIME it has taken. You cannot simply roll distance and time into one. They are very different and they must be used separately.

We can observe a distant galaxy behind us, approx 13.8 billion light–years away. We can also observe a galaxy ahead of us by about the same distance. Remember that the measurement of light–years is a distance across the universe, not the number of years of travel.

We must also place *the young galaxy* on the other side of the–beginning–of–time, or the calculations could never work. For example we cannot observe other galaxies at that age, because the light from (then) has passed us by a very, very long time ago.

There are millions of galaxies out there, so what about seeing galaxies ahead of us? Some galaxies ahead of us also **appear** to be around 13 billion light years away. But the difficulty lies in trying to get the galaxies in place and on time.

Light–years is not a measure of time, it is a measure of distance. The light travels at a rate of trillions of kilometers per year. Compared to (580 * 60 * 60 * 24 * 365) kilometers per year for our galaxy, which is approximately 18 290 880 000 kilometers per year.

Light travels at approx 299 792 kilometers per second.

Each year the light travels 9.4 trillion kilometers.

From the time when our Sun 'lit–up' (approx 4.5 billion years ago) this sytem has only travelled through space a distance of… (8.2309E+19) kilometers or approximately 90 million light–years. Notice that there is a vast difference between the **time** (4.5 billion years) and the **distance** (9 million light–years). With these figures in mind you can see how difficult it is, to get all the galaxies in the right place in time for us to see the light arriving from them. Remember that the light is coming **from where those galaxies were** 13 billion years ago, **not from where they are now.**

As for the size of the universe, it must be greater than 80 billion light years across. This can be determined by the fact that we can only see light from the systems in the past. We cannot see them where they are now because the light from there hasn't arrived in out part of the universe, and in many cases it will be another 13 billion years before we could begin to see their (present) light arriving.

The far off galaxies ahead of us MUST be travelling at less than the speed of light or we couldn't see them at all. Even if they are getting along at a cracking pace, the universe must be at least 60 billion years old, but I suspect that it is much more than that.

Regardless of how you process the age of the universe, a proper study of evolution must start in an empty space (without matter).

For example if there were already many 'suns' in existence before time began, then the process could revert to chaos. If there was any matter in existence before the big–bang occurred it could have spread the new material in a random and scattered way, and it would be very unlikely to form a system or a planet. It would be very unlikely to become stable. So does any of the above (rule–out) a previous universe?

No, not at all, it just rules out the chance of us seeing such a thing (because the light from 'then' has already passed us by, billions of years in the past).

Our Sun is made up mostly of hydrogen and helium. At the Sun's core, hydrogen is converted into helium in a nuclear reaction, releasing energy. In the process, (when two protons collide), one of the protons changes into a neutron. The two bond together, forming a particle known as a deuteron, (constructed of one proton and one neutron).

The formation of deuterons in the Sun's core would never take place if the neutron mass was significantly greater or less–than the proton. Deuterons could not form unless the relative mass of each particle was within 0.1 percent.

Stars are able to produce nuclear energy through the formation of deuterons. Without this critical process, no star would produce enough sustained energy to support life on any planet orbiting around it. Deuterons are vital to sustaining the Sun's thermonuclear reaction, which provides sufficient energy to sustain life on earth.

The life of a neutron outside the nucleus is about 15 minutes. It decays into a proton and an electron. If a neutron were only a tiny fraction smaller, free protons (particles that are not part of a nucleus) would then decay into neutrons—and atoms simply would not exist! In such a case, free protons would decay into neutrons, and—since the nucleus of a hydrogen–atom is simply a free proton—hydrogen could not exist!

Thus, a relative change of even the slightest proportions between neutrons and protons would eliminate hydrogen, the most abundant element in the universe. Without hydrogen—**water** (H^2O)—the basic solvent for all biological life—**would not exist**. In short, if the relative mass of proton to neutron deviated beyond one part in 1000—life could not exist!

Proton Charge equal to Electron Charge

Scientists have been able to measure and compare the relative proton and electron charge within atoms, and have established that these charges can only differ by less than one part in 1.00E+15 Therefore, since the charge of the electron is of equal magnitude to the charge of the proton, atoms tend to maintain a neutral charge.

However, if one of these charged particles differed by a tiny amount then an atom would no longer be electrically neutral. If the proton charge were greater, atoms would be electrically positive. If the electron charge were greater, then atoms would become electrically negative. In such cases, atoms would no longer be neutral, but would possess a definite charge—positive or negative. Since like–charges repel, there would be repulsion between atoms of elements and solid matter could not exist!

Strong Nuclear Force

The force that binds the particles of the atomic nucleus together is called the strong nuclear force. If the strong nuclear force were only about three percent stronger, then all the hydrogen in the universe would have long since been transformed into helium! Such an increased force would cause two protons to form a helium nucleus lacking a neutron. Since the strong nuclear force is not quite strong enough to bring about this reaction.

We have hydrogen in abundance, so vital for an environment favourable for life—providing for the existence of water, and energy for the Sun. Stars fueled exclusively by helium would be short-lived, and would be unstable during their formation process.

The strong nuclear force has to fall within a relatively narrow band for a balanced universe (suited to sustaining life).

Epsilon Constant – Gravitational Fine Structure

Concerning the universe, if the epsilon constant (relating to gravitational forces) deviated only slightly (in one direction) in relation to gravitational fine structure, all stars would be 'red dwarfs'. (A Dwarf star in our time—generally a white dwarf—is the remaining core of a star that has nearly finished its life–cycle.

If the epsilon constant deviated (in the other direction), all stars would intensify into blue giants—huge stars with energy levels of enormous intensity. As an example, of two stars in the neighborhood of our Sun, Rigel a blue giant, is over five times hotter than Betelgeuse, a red super–giant in the latter stages of its life, (the red giant eventually collapses into a white dwarf).

Defining the above forces is beyond my ability however it is sufficient to say that they are critical for the stability of the universe. The epsilon constant is defined as the fine structure constant to the twelfth power, when multiplied by the electron/proton mass to the fourth power. (The value of the epsilon constant is a tiny value).

This is an extremely delicate force that must not have the slightest deviation. The value of the gravitational–fine–structure, relative to the epsilon–constant, is equally critical for stability. On a finely calibrated instrument one kilometer long, the tolerance (for the range of this force) could be no wider than one millimeter.

Scientists agree that neither a blue giant nor a red dwarf can support life on an orbiting planet. The exact balance described **is** required for biological life to exist. The slightest deviation in one direction or the other would cause all the stars in the universe to quickly develop into either blue giants or red dwarfs.

As discussed earlier, the expanding–universe was held–back, by the force of gravity. The hold of gravity decreases with distance. Imagine a force with the opposite behaviour so that it increases with distance. This opposite force would be the cosmological constant.

The value of the cosmological constant is very close to zero. If the level increased, a distortion of space–time would take place over shorter distances, in unsuitable conditions planets would not be able to have suitable orbits around stars (because of the distortion of space–time). We don't try to understand this concept, but to consider the precision of the cosmological constant, in order for the universe to produce a life–supporting planet.

A different cosmological constant would mean a different rate of expansion. The rate of expansion that did occur, allowed for the formation of the vast number of stars in millions of galaxies. A faster rate would have *prevented* the formation of stars. A slower rate would have caused matter to form into black holes. (At the time of writing 'black holes' are thought to result from the collapse of massive stars). The one correct level of the cosmological constant is a tiny number that approaches the value of zero. This is a very sensitive and intricate force and it had to stabilize at an optimal value (close to zero) in order for the universe to form.

A weak nuclear force is needed:

The weak nuclear force allows a proton to change into a neutron at the optimum rate. If this force were only slightly smaller, then all of the hydrogen in the universe would have long since been changed into helium. As covered earlier, hydrogen is an essential part of the water molecule—and water is essential for all biological life.

Two types of reactions take place in a star (in the production of energy). The first is the formation of a deuteron 9 as two protons collide, producing one proton and one neutron bound together.

The second reaction occurs when a deuteron collides with a proton, producing a helium nucleus with an emission of energy. Unless the weak nuclear force existed at a very special level, deuterons would never form in the first place. A steady rate of reaction—caused by the strong nuclear force relative to the weak nuclear force—is what allows the Sun's thermonuclear reaction to be maintained.

Some of the information in this article was summarized from a program known as Universe Analyzer.

The software demonstrates how remote the probabilities were for all of these requirements to be met. One of the demonstrations features over 2000 separate universe models. These models give a realistic picture of what the **chance of success** would be in real terms.

Not a single experimental model was able to meet three or more of the requirements for success. The user could alter the parameters to 'tweak' the forces and constants found in the 'universe' in the hope of scoring a higher probability than the demonstration.

Given the constant values, the forces described and other parameters, the probability of these requirements being met by chance would be effectively ZERO.

Once evolution is allowed the existence of an orderly universe, (favourable to organic life), then the magic of abiogenesis has a nice cushy place to occur, but the existence of an observable universe, (with an observer–species living on one of its planets), should never be taken for granted.

Chapter Eleven

USE YOUR OWN LOGIC

The Rose of Jerico:

This is a plant that lays dormant in the desert and bursts into life as soon as it rains. The reaction to rain is a pre–programmed response.

A **process** needed to be in place right from the get–go. When you apply the theory of slow gradual change to this plant you must also apply a "time before." There had to be a time in the development of this species when the pre–programmed response was not yet in place (for slow and gradual change to occur over time).

Without the pre–programmed response the plant would never wake up. It would simply stay dormant until it dies – every single time. Once they have all died off the species would be extinct, yet it is alive and well. The existence of this specimen is evidence against scientific evolution.

Oil birds:

This creature is a burden to the culture of scientific evolution. It is a unique bird that lives in the caves of Trinidad and Tobago. This bird would be a real headache to try and fit into the scientific tree of life. It is a vegetarian—nocturnal bird. It lives on a fruit diet, yet it can fly over 200 kilometers at night when needed.

It can navigate in a PITCH DARK CAVE (using echo location). This survival mechanism requires the baby bird to drop from its ledge (in the pitch dark) and fly away. There is only one way this bird could have survived such a life. It had to have all the right abilities from the start. If you apply the slow and steady approach to this species it just wouldn't make it. With the gradual change model there would have to be a time before this bird had the ability to send and receive signals in the dark. During the evolution stage each and every new chick would have smashed its way to the bottom of the cave and died. The creatures that die cannot pass–on any improvement to the next generation.

The crocodile:

Yes even the crocodile smiles at evolution!

Scientists believe that the tooth anatomy of the crocodile has not changed in the past 200 million years. That's just a bit much (slow – and – steady) don't you think?

The bombardier beetle:

This beetle contains hydroquinol AND hydrogen–peroxide in tiny chambers inside its body. These two things *can* be stable when mixed together, but if you introduce some heat it will shoot out across the lab.

The beetle doesn't need to use heat, it introduces a catalyst into the hydrogen peroxide and another enzyme and this causes the two products to react at a lower temperature. The output from this process can be directed to the front or to the rear of the creature. When activated, the reaction takes place hundreds of times per second and produces a noise and fumes to repel an enemy.

Try to imagine a way that this beetle could develop this ability over time by a process of random change. Such a thought should boggle your mind!

There are two possible outcomes for the experimental model…

- The bombardier mechanism misfires because of an unfinished part, killing the beetle, or
- The predator catches and kills the beetle.

Either of the above will end the evolutionary process, long before the product is finished (and ready for survival).

Sharks:

Sharks are known to have poor eyesight. They make up for this by having two other senses. One is the ability to sense electrical impulses from their food source. This feature could not be produced by gradual change. The pre–shark needed food energy to build the sensors. Yet it needed the sensors to locate the food source. The pre–shark would have died of starvation, thus leaving no offspring.

Snakes:

Snakes survive by crushing and then swallowing their food. The eating process employs a breathing tube. The pre–snake would have required vast amounts of food energy to produce a breathing tube. Without the breathing tube in place, the pre–snake would have died from lack of oxygen every time it tried to eat, thus leaving no offspring.

Consider what we know about computer coding. In the early days of computers I would write computer programs in a language called BASIC.

The program once completed is run and tested. While the program is running there is an opportunity to input some data then press enter and see a result.

After I have seen the output from the code I might decide to make some improvements. For example—to display a list of the values entered, as well as the final result. At this point a new piece of code is required. It is impossible to add the new code without providing a link from the old code to the new. This 'improvement' always involves a change to the existing code and it always requires something to be added at the correct point in the existing code.

You can be sure that the same problem would exist when tinkering with modulo four coding. The new module (generated by evolution) could not build anything unless it was somehow linked into the existing code. This would need to be done in a very precise way. Random changes over vast periods of time cannot produce this one correct change. The new code would simply be ignored until it was somehow linked into the existing code in a very precise way.

When humans use 'bots' to make a hole in a cell and modify genes, they rely on the internal mechanisms to complete the task, otherwise their attempts at modification would fail.

When you begin with something that works and you randomly modify it, it will get worse not better. The process is known as genetic entropy. The scientific theory of evolution requires such a modification to occur in the opposite direction.

The problem for random evolution is that there are always more disordered states than there are ordered states. If you make random changes over vast periods of time the outcomes must be more disorderly. The idea of improving something by random change will always be nothing more than wishful thinking. There should be no place in science for wishful thinking to flourish.

The development of the first living cell is the biggest problem for evolution to overcome. The first evoluture had to arrive by a process of self–organisation. Such a thing has no reason to be reproductive. It had no reason to think about the process that occurred—it had no reason to believe that such a process would stop occurring. Evolutionists may study the duplicating process, yet they have not found a mechanism that would *invoke* such a process in the first place. The start–up of the duplication and reproduction process would have to be very carefully timed. The process of coming–about–from–non–living would need to wind down at the same time as the duplication process was perfected. In other words, how would the evolutures know when to stop building themselves from scratch and start reproducing?

Smoke and mirrors:

If you consider the evidence for the first living cell **with the untrained eye**, you may be able to spot some of the problems.

The science promoter will use snappy terms like primordial–ocean.

But – we're not talking about an ocean are we?

- We're not talking about a drop in the ocean.
- We're *ARE* talking about a microscopic cell **smaller than a drop of water**.

If such a process could easily start–up (on its own) at a random time in history, what caused it to stop? Why doesn't this process continue? Something that is incredibly easy, something that is self motivating – yet it only happened for a brief moment in the life of a planet, sounds a bit contradictory to me!

The Australian termite:

Biologists struggle with the existence of this creature.

This little-known termite differs from any others. It is four–things–in–one. Each depends on the others for their life, a case in which you cannot have one without the others. Consider this a curiosity:

In microbiology; some students have observed a *microorganism* called *Mixotricha Paradoxa*, a survival mechanism for Australian termites. When it was first *discovered*, it looked like it was covered with curly hairs. Looking closer, it was revealed that they were not hairs, but *spirochetes*, a totally different type of microorganism. On the *Mixotricha*, there were bumps [or appendages] where the spirochetes attached, and *bacillus* which lodged on the other side of the bump. The spirochetes provided a means of locomotion for the entire colony (of microorganisms). These *three totally different germs just decided to live together* (in a community).

So what? Well if the math is difficult for two beings, here is an inter–dependence between a large microorganism, a spirochete, a bacillus, and an Australian termite. And if you take it one step further—none of these could survive unless the building parts in the tree provided the right "food."

I suppose for the evolutionist, you could teach that at one point in time they formed a committee and decided to work together; the *Mixotricha* 'developing' bumps for the spirochetes to utilize. Meanwhile providing a place for the bacillus to hide, then they all just 'decided' to live in the food supply system of a termite.

That all sounds very cushy and very easy, (except for the mathematical calculations). What is the chance for all three products, each becoming a "finished product," at the same time? They must finish *evolving* just in time for the termite to survive as a finished–product. They also needed to *survive* inside the termite–gut as their evolution progressed from simple to complex. While you *imagine* these changes, remember that the chlorella didn't '*evolve*' at all, for 540 million years.

Is it ok to say that *they* could **NOT** possibly have developed separately—by a slow and steady process of (*random*) change?

In our man–made technology the most outstanding thing is that every device becomes obsolete within a year or two. Some expensive items (such as game consoles) become obsolete before they're even unpacked from the box. You become familiar with this cycle of early death. You might say it is taken for granted. The products that you buy are carefully planned designed and manufactured by several tightly controlled processes, yet they are soon out of date, they soon become useless.

The processes that build a human body from the inside out are thought to be random changes over time.

Yet how could a random process build a piece of technology that is far superior to the one that was planned and designed and built by humans? A superior process that rebuilds parts on the fly!

For the logical mind the technology gap is staggering, even impossible. Our own human (internal) technology hasn't needed an upgrade for thousands of years.

Sure people complain about the decline in quality of their body cells, but in almost every case the decline is self inflicted. If you look after your health and breathe some fresh air your body will keep itself in fine condition for a very long time.

It's very lucky for us that nature didn't learn things the way that humans do. If nature started planning and designing and building living things in the way that humans do, we would be in deep shit very quickly. Imagine if OUR operating system was thrown together by humans. The blue–screen–of–death would be a literal death every time. There is no reset button there is no re-installing the operating system, (just to keep it half working). Lucky for us there is no startup repair program in our internal code. Every computer I have owned (over the past ten years or so) has (at some point) gone into the startup repair mode. None have ever recovered by that method. The Microsoft–corporation would look more professional if they removed "startup repair" from their systems. The only thing worse than a full system failure is its failure to know what's wrong, coupled with a failure to repair anything, ever. How embarrassing?

If the internal workings of the human body look complex, then they must be complex. Scientists have given themselves the right to teach us that something horrendously complex came about by blind chance over vast periods of time. That particular 'scientific' method–of–operation is an insult to my intelligence.

From the horses' mouth:

Horticultural experiments have established, beyond the possibility of refutation, that botanical species of each genus appeared—in an original and highly plastic condition, and that these have produced—chiefly by intercrossing, but likewise by variation, **all our existing species.**

[The Dean of Manchester, "Horticultural Transactions" Published in 1822 and again in 1837]

Chapter Twelve

THE SEVEN YEAR CYCLE

There are two distinct ways to study evolution.

One is to read the science and blindly accept everything you find. The scientists must be smart enough to know what they're doing, right?

Another method is to think it through.

In this chapter— try to think through the known information regarding the scientific theory of evolution.

Every part of every creature has to be considered from an evolutionary point of view. The process of evolution must not start with a completed creature. If something seems too easy or too convenient just take it back to a point in the past. This is a very easy procedure for the logical mind.

Starting from the human, work back to the monkey's cousin then the pre–monkey's cousin (and so on), let your mind drift further and further back in time. Eventually you will arrive at a living thing that doesn't have the abilities that we have.

The most difficult problem for the hardened evolutionist is the seven–year cycle of life. For evolution to be true there must have been a distant relative (a lesser species) that did not have this cycle of tearing down and rebuilding. So what are the implications?

There was a time when each and every living thing had less than seven years to live. Along with that comes another problem – it had less than seven years to figure out that it was dying too young.

There was (according to evolutionary theory) a time before this re–building could occur. There was a time when every living thing would die in less than seven years. Most likely in just a few days!

Right there is a whole new problem for the beginning of the process (of human reproduction). It takes more than seven years for a human to develop to maturity. Without the ability to tear down and re–build living parts (while they are in use) the human system (or the pre–human system) could **not** possibly make it past the first generation. The same failure would occur over and over again as the new evolutures were 'pumped out by their predecessor.'

The process of rebuilding is enormously complex and highly controlled. Yet while ever we are able to supply the correct parts the re–building process will continue for years on end.

Something in the past, somewhere between the first living cell and the pre–monkey's cousin had to anticipate the need for rebuilding. Blind random changes cannot anticipate a future need. The idea of a simple living thing developing a system that would allow it to live longer, goes way beyond the scope of random evolution.

Sometimes when something sounds impossible it really is impossible. The word itself is not something you can manipulate to get a better outcome. If the facts of evolution point to the word impossible then you should start thinking about what impossible means. Scientists are very clever, but they cannot observe this impossible event. Science is a testing and proving occupation. Science is not in the business of relying on assumptions to hold onto a belief; except when it comes to the scientific theory of evolution.

Scientists are able to observe the processes of tearing down and rebuilding (the entire body). They believe that evolution 'came up' with this process, yet they are unable to find a way that the process could actually have started in the first instance.

The process itself is there for us to discover and observe, but the process that started the process is nowhere to be found.

Are you able to deal with that unlikely event? Are you able to believe that some living–thing in the past, could manipulate its inner workings, so that it could now seek–out dying parts and tear them down, and then rebuild them almost as good as the original.

Even Maxwell Smart would find that hard to believe!

Would you believe that there are 22 different spare parts that can be used to manufacture 250 different protein–shapes? Would you believe that each different shape has a specific purpose? Would you believe that each of the manufactured parts had to evolve by a random–process? Would you believe that many thousands of different products developed the same parts in the same timeframe? Would you believe in nature sprites?

I don't think so!

Common ground:

In my first book I have provided information that clearly shows the method–of–operation used by the fundamental religions of this world.

The false religions claim to be following the Author of a particular book. However the only way they can continue their man–made belief is to take the marker–pen to the words written by the original Author! They must blacken out the parts that don't fit with their inventions. They must disobey the very thing that they claim to follow!

In this second book I have provided information that clearly shows the method–of–operation used by the fundamental evolutionists of this present world.

The false Darwinist claims to be following the author of a particular book. However the only way they can continue their man–made belief is to take the marker–pen to the words written by the original author! They must blacken out the parts that don't fit with their own inventions. They must disobey the very thing that they claim to follow!

The last word:

> In my introductory book "Artificial Religion"
> I included a brief story about...

...The 16th century Italian philosopher (and former Catholic priest) Giordano Bruno who was charged and then killed—for a stubborn adherence to his then **unorthodox beliefs**—including the idea of the universe as infinite space, and the idea that 'other solar systems' exist in the universe...

The most important point I could ever make is wrapped up in Giordano's untimely death. The popularity and peer pressure that ruled the best minds of that era was all about (the theory of all things geocentric). Giordano Bruno was playing a lone–hand whenever he spoke to others about his belief. The popular majority shut him down, every time he opened his mouth to explain. Just as the popular majority would shut me down every time I open my mouth to explain the data.

I'd like YOU (as a fellow human being) to consider those times very carefully. The "popularity" was opposed to the view that the Sun is best described as the center of this solar system. Keep in mind always that the combined wisdom of all his opponents wasn't enough to make the geocentric idea a reality.

Their belief was very strong and they shouted him down every time he tried to explain. Yet Giordano Bruno knew more about the universe than all those minds put together.

The scientific establishment has learned nothing from the mistakes of the past. They still believe (with all their heart) that being popular, somehow translates to being correct?

Without resorting to any deception of any kind, we can find the truth about the invention of new words. Abiogenesis is not a scientific theory, **it is a hypothetical phenomenon.**

If you stick to the rules of logic and apply the known facts, you can find a logical outcome. The first occurrence of macroevolution required life–from–non–life.

Macro is supposedly lots of micro, over a very–very long period of time. Life from non life is a hypothetical event from the scientific mind.

"The Scientific Establishment" manages to roll together two very different types of evolution (by saying that one naturally follows the other). The real problem (for the man–kind) is that the one that supposedly follows the other (the macroevolution) must in reality be the very first event (therefore it **must be a hypothetical event**).

If you look at the evidence for the first living cell (with the untrained eye) it becomes quite remarkable). Not an ocean of sea water, but a microscopic part of one drop of water.

It is very easy for the dedicated scientist to believe these things.

In that tiny space, the entire gene sequence had to gather together, with all the amino acids in the correct order. (Some of the information that is now common to plants, animals and man).

For them—all those parts got together in the exact sequence, including the ability to join the parts needed to make the first strand. The strand also included the ability to utilize its own information to make copies of itself (without changes to the pattern). Then, all of a sudden, evolution stopped making random changes and kept the good–knowledge so that it spread throughout the plant and animal domains for millions of years.

Perhaps nature–sprites are more powerful than I thought?

High school science teaches that, building–blocks (or anything else) in water will revert to a state of chaos. This was demonstrated in class by adding a 'tint' to a clear beaker of water. At first I could see a long streak, slowly descending through the water. By the end of the lesson the beaker was a uniform colour throughout. Yet, science teaches that the parts for the first living cell were able to defy this process and gather together in an orderly fashion?

Let the scientist teach that macro–evolution has never been observed… by excluding the dodgy experiments that are designed to deceive us.

Let the scientific establishment admit that they do not accept the work of Charles Darwin.

When these two things happen, the man–kind will have two good reasons to begin trusting the scientific 'method of operation' once again.

The End

Printed in the United States
by Baker & Taylor

Printed in the United States
By Bookmasters